The Empathetic Brain: From Neurons to Social Change

Azhar ul Haque Sario

Copyright

First Printing, 2024

Azhar.sario@hotmail.co.uk

ORCID: https://orcid.org/0009-0004-8629-830X
Disclaimer: This book is free from AI use. The cover was designed in Microsoft Publisher

Contents

The Neuroscience of Empathy

1.1 The Mirror Neuron System and its Role in Social Cognition

The discovery of mirror neurons in the 1990s revolutionized our understanding of social cognition. These specialized neurons, located in the premotor cortex and inferior parietal lobule of the brain, fire both when we perform an action and when we observe someone else performing the same action. This "mirroring" mechanism allows us to understand the intentions, emotions, and experiences of others, forming the foundation for empathy and social connection.

The Discovery and Location of Mirror Neurons

Mirror neurons were first identified in macaque monkeys by Giacomo Rizzolatti and his colleagues at the University of Parma. While studying neurons involved in hand and mouth movements, they noticed that some neurons fired not only when the monkey grasped an object but also when it observed a researcher grasping the same object. This unexpected finding sparked a wave of research exploring the role of mirror neurons in primates, including humans.

In humans, mirror neurons have been found in the premotor cortex, a region involved in planning and executing movements, and the inferior parietal lobule, which integrates sensory information and plays a role in spatial awareness and attention. These areas are interconnected and form part of a larger network involved in action understanding and social cognition.

How Mirror Neurons Work

The "mirroring" function of these neurons is thought to arise from their ability to simulate the observed action in the observer's own motor system. When we see someone grasp a cup, for example, the mirror neurons in our brain fire as if we were performing the action ourselves. This simulation allows us to understand the goal of the action (to pick up the cup) and the intention behind it (to drink from it).

This process is not limited to simple motor actions. Mirror neurons also respond to facial expressions, allowing us to understand the emotions of others. When we see someone smile, the mirror neurons in our brain associated with smiling fire, creating a sense of shared feeling and facilitating emotional connection.

Implications for Social Cognition

The discovery of mirror neurons has profound implications for our understanding of social cognition. By allowing us to simulate the actions and emotions of others, mirror neurons provide a neural basis for:

Understanding intentions: We can infer the goals and intentions behind others' actions by simulating them in our own motor system.
Sharing emotions: We can experience the emotions of others through the activation of our own emotional circuits.
Developing a sense of shared experience: By mirroring the actions and emotions of others, we create a sense of shared experience and connection.

Examples of Mirror Neuron Activity

Numerous studies using functional magnetic resonance imaging (fMRI) have demonstrated the activity of mirror neurons in humans. For example:

Facial expressions: When participants observe facial expressions of emotions, such as happiness, sadness, or

anger, the mirror neuron system in their brain is activated, particularly in the inferior frontal gyrus and the insula.
Hand gestures: Observing hand gestures, such as pointing or grasping, also activates the mirror neuron system, suggesting that these neurons are involved in understanding the communicative intent behind gestures.
Goal-directed actions: Studies have shown that mirror neurons fire more strongly when observing actions that have a clear goal, such as reaching for a cup to drink, compared to actions that are less goal-directed. This suggests that mirror neurons are not simply responding to movement but are also involved in understanding the purpose behind actions.

Mirror Neurons and Autism Spectrum Disorder

Research suggests that individuals with autism spectrum disorder (ASD) may have impaired mirror neuron function. This impairment may contribute to the social difficulties characteristic of ASD, such as difficulty understanding social cues, interpreting emotions, and engaging in social interactions.

Studies have shown reduced activity in the mirror neuron system in individuals with ASD when observing facial expressions and imitating actions. This reduced activity may explain why individuals with ASD often have difficulty understanding the emotions and intentions of others, leading to challenges in social communication and interaction.

Recent Advances and Future Directions

In 2025, research on mirror neurons continues to expand our understanding of their role in social cognition and their implications for various conditions. Some of the latest findings include:

Refinement of Mirror Neuron Theory: Researchers are refining the original mirror neuron theory to incorporate

new findings and address criticisms. Some argue that mirror neurons may not be a distinct class of neurons but rather a network of neurons with diverse functions.
Role in Language Development: Studies suggest that mirror neurons may play a role in language development by facilitating the imitation of sounds and gestures, which are essential for language acquisition.
Therapeutic Applications: Researchers are exploring the potential of using mirror therapy, a technique that uses mirrors to create the illusion of movement, to rehabilitate individuals with stroke and other neurological conditions. Mirror therapy may activate the mirror neuron system and promote motor recovery.
Expert Insights:

Dr. Vilayanur Ramachandran, a renowned neuroscientist, suggests that mirror neurons may be the "neurons that shaped civilization" by enabling the development of culture, language, and empathy.
Dr. Marco Iacoboni, a leading researcher in the field, emphasizes the role of mirror neurons in social learning and imitation, arguing that they allow us to "learn by feeling."
Conclusion:

The discovery of mirror neurons has provided a crucial link between brain activity and social cognition. By allowing us to simulate the actions and emotions of others, mirror neurons enable us to understand intentions, share emotions, and develop a sense of shared experience. Further research on mirror neurons promises to deepen our understanding of social interaction, empathy, and the human condition.

1.2 Emotional Contagion: The Neural Mechanisms of Shared Feeling

Emotional contagion is a fascinating phenomenon where emotions, like laughter or fear, seem to spread from one person to another. This "transmission" of emotions occurs through subtle cues like facial expressions, vocal tone, and body language. The neural mechanisms underlying emotional contagion involve the limbic system, particularly the amygdala and insula, and the autonomic nervous system.

The Limbic System and Emotional Contagion

The limbic system is a collection of brain structures involved in processing emotions, memory, and motivation. Two key structures within the limbic system play a crucial role in emotional contagion:

Amygdala: This almond-shaped structure is responsible for processing emotions, particularly fear and anger. It also plays a role in recognizing emotions in others. When we observe someone expressing fear, our amygdala activates, triggering a fear response in ourselves.

Insula: This region is involved in processing a wide range of emotions, including disgust, happiness, and sadness. It also plays a role in interoception, the awareness of our own bodily sensations. The insula helps us to connect our own emotional experiences with the emotions we observe in others.

The Autonomic Nervous System and Emotional Contagion

The autonomic nervous system (ANS) regulates involuntary bodily functions, such as heart rate, breathing, and sweating. It plays a crucial role in emotional contagion by mediating the physiological responses associated with emotions. When we observe someone expressing an emotion, our ANS responds by triggering physiological changes that mirror the observed emotion. For example, if we see someone who is anxious, our heart rate may increase and we may start to sweat.

How Emotional Contagion Works
Emotional contagion is thought to occur through a combination of:

Facial mimicry: We unconsciously mimic the facial expressions of others, which activates the same emotional circuits in our brain.
Vocal tone and body language: We also pick up on subtle cues in vocal tone and body language, which convey emotional information.
Empathy: Our ability to understand and share the feelings of others allows us to experience their emotions as our own.
Examples of Emotional Contagion
Numerous studies have demonstrated the phenomenon of emotional contagion. For example:

Physiological responses: Studies measuring physiological responses, such as heart rate and skin conductance, have shown that individuals observing emotional expressions in others exhibit similar physiological responses. For instance, observing someone who is stressed can lead to an increase in our own stress hormones.
Mood convergence: Research has shown that spending time with people who are happy or sad can influence our own mood. This mood convergence is particularly strong in close relationships.
Social media: Emotional contagion can also occur online. Studies have shown that exposure to positive or negative posts on social media can influence our own emotions and behavior.
Emotional Contagion and Empathy
Emotional contagion is considered a basic form of empathy, as it allows us to share the emotions of others.

This shared feeling can promote social bonding and cooperation. However, emotional contagion can also have negative consequences, particularly for individuals in certain professions.

Emotional Contagion in Healthcare and Social Work
Healthcare professionals and social workers are often exposed to intense emotions, such as sadness, fear, and anger. This exposure can lead to emotional contagion, causing these individuals to experience emotional exhaustion, burnout, and even secondary traumatic stress.

Strategies to mitigate the negative effects of emotional contagion in these professions include:

Self-awareness: Recognizing one's own emotional responses and triggers.
Emotional regulation: Developing coping mechanisms to manage emotions.
Mindfulness: Practicing mindfulness to stay present and grounded.
Social support: Seeking support from colleagues and supervisors.
Recent Advances and Future Directions
Research in 2025 continues to explore the complexities of emotional contagion and its implications for various fields. Some of the latest developments include:

Neural underpinnings: Advanced neuroimaging techniques are providing a more detailed understanding of the brain regions and networks involved in emotional contagion.
Individual differences: Researchers are investigating individual differences in susceptibility to emotional contagion, exploring factors such as personality traits, genetics, and social experiences.

Artificial intelligence: AI is being used to detect and analyze emotional cues in various contexts, such as customer service interactions and social media.
Expert Insights:

Dr. Elaine Hatfield, a pioneer in the study of emotional contagion, emphasizes the role of nonverbal cues and social context in shaping emotional experiences.
Dr. Sonja Lyubomirsky, a leading researcher in positive psychology, highlights the importance of managing emotional contagion to promote well-being.
Conclusion:

Emotional contagion is a powerful force that shapes our social interactions and emotional experiences. By understanding the neural mechanisms and consequences of emotional contagion, we can develop strategies to harness its positive effects and mitigate its negative impact, particularly in professions that involve high levels of emotional exposure.

1.3 The Default Mode Network and its Contribution to Empathy

The Default Mode Network (DMN) is a large-scale brain network that has garnered significant attention in recent years for its role in social cognition, self-awareness, and theory of mind. This network, active when we are not focused on the external world and seemingly "at rest," plays a crucial role in our ability to understand others' mental states and perspectives, contributing significantly to our capacity for empathy.

What is the Default Mode Network?
The DMN is a collection of interconnected brain regions that show increased activity when we are not engaged in goal-directed tasks. This network includes areas such as:

Medial prefrontal cortex (mPFC): Involved in self-referential processing, social cognition, and theory of mind.
Posterior cingulate cortex (PCC): Plays a role in autobiographical memory, self-awareness, and emotional processing.
Angular gyrus: Involved in language processing, semantic memory, and social cognition.
Hippocampus: Critical for memory formation and retrieval, including social memories.
These regions work together to support a range of cognitive functions, including:

Mind-wandering: Spontaneous thoughts and mental imagery.
Self-reflection: Thinking about oneself, one's past, and one's future.
Social cognition: Thinking about others, their thoughts, and their feelings.
Perspective-taking: Understanding the viewpoints of others.
The DMN and Social Cognition
The DMN is increasingly recognized for its crucial role in social cognition. When we engage in social interactions, our DMN helps us to:

Make sense of social cues: Interpreting facial expressions, body language, and vocal tone.
Infer the mental states of others: Understanding their thoughts, beliefs, and intentions.
Predict their behavior: Anticipating how others might act in different situations.
Empathize with others: Sharing their feelings and perspectives.
The DMN and Theory of Mind
Theory of mind (ToM) is the ability to understand that others have their own thoughts, beliefs, and desires that may differ from our own. This ability is essential for social

interaction and empathy. The DMN plays a key role in ToM by:

Simulating mental states: Creating mental models of others' minds.
Perspective-taking: Imagining the world from another person's point of view.
Understanding social narratives: Comprehending stories and situations that involve social interactions.
Examples of DMN Activity in Empathy
Brain imaging studies have revealed increased DMN activity during tasks that involve thinking about oneself and others, such as:

Autobiographical memory retrieval: Recalling personal experiences, which often involve social interactions and emotions.
Perspective-taking tasks: Considering a situation from another person's point of view.
Moral dilemmas: Making judgments about social situations that involve ethical considerations.
Empathy tasks: Responding to scenarios that evoke empathy, such as witnessing someone in pain or distress.
Disruptions in the DMN and Social Difficulties
Disruptions in the DMN have been linked to social difficulties in various conditions, including:

Schizophrenia: Individuals with schizophrenia often exhibit reduced DMN connectivity, which may contribute to their difficulties with social cognition, ToM, and empathy.
Autism spectrum disorder: Some studies suggest altered DMN activity in individuals with ASD, which may be associated with challenges in social interaction and understanding social cues.
Depression: Depression has been linked to increased DMN activity, which may contribute to rumination and social withdrawal.

Recent Advances and Future Directions (2025)
Research on the DMN is rapidly evolving, with new findings emerging in 2025 that shed further light on its role in empathy and social cognition:

Dynamic nature of the DMN: Studies are revealing that the DMN is not a static network but rather a dynamic system that can flexibly reconfigure its activity depending on the task demands.
Individual differences in DMN activity: Researchers are exploring how individual differences in DMN activity relate to personality traits, social skills, and susceptibility to mental health conditions.
Training the DMN: Interventions such as mindfulness meditation and compassion training have been shown to modulate DMN activity, potentially enhancing empathy and social cognition.
Expert Insights:

Dr. Randy Buckner, a leading researcher on the DMN, emphasizes its role in constructing mental simulations and understanding social narratives.
Dr. Matthew Lieberman, a social cognitive neuroscientist, highlights the importance of the DMN in social connection and its potential to be strengthened through social interaction.
Conclusion:

The DMN plays a crucial role in our ability to understand ourselves and others. By supporting self-reflection, social cognition, and theory of mind, the DMN enables us to navigate the social world and connect with others on an emotional level. Further research on the DMN promises to deepen our understanding of empathy, social behavior, and the human mind.

1.4 The Prefrontal Cortex: Regulation, Cognition, and Moral Judgment in Empathy

The prefrontal cortex (PFC), often referred to as the brain's executive control center, plays a critical role in empathy by regulating emotional responses, enabling cognitive empathy (perspective-taking and mentalizing), and guiding moral judgments. It integrates emotional and cognitive information to shape our empathetic behavior and social decision-making.

The Prefrontal Cortex and Emotional Regulation

The PFC is crucial for regulating emotions, a key component of empathy. It helps us to:

Inhibit impulsive reactions: When we encounter someone in distress, the PFC helps us to control our initial emotional responses and avoid acting impulsively.
Manage emotional intensity: The PFC modulates the intensity of our emotions, preventing us from becoming overwhelmed by the emotions of others.
Reframe emotional experiences: The PFC allows us to reappraise emotional situations, shifting our perspective and reducing negative emotional impact.
This regulatory function of the PFC is essential for maintaining emotional balance and engaging in empathetic behavior effectively.

The PFC and Cognitive Empathy

Cognitive empathy, also known as perspective-taking or mentalizing, involves understanding the mental states of others, including their thoughts, beliefs, and intentions. The PFC plays a central role in cognitive empathy by:

Inferring mental states: The PFC helps us to deduce the mental states of others based on their behavior and social cues.

Taking different perspectives: The PFC allows us to imagine the world from another person's point of view, understanding their unique experiences and challenges.
Predicting behavior: The PFC helps us to anticipate how others might think and act based on their mental states.
This cognitive aspect of empathy allows us to connect with others on an intellectual level, fostering understanding and compassion.

The PFC and Moral Judgment

Empathy is closely linked to moral judgment, as our ability to understand and share the feelings of others influences our ethical decisions and actions. The PFC plays a key role in moral judgment by:

Evaluating social situations: The PFC helps us to assess social situations and identify the ethical implications of different actions.
Weighing moral considerations: The PFC allows us to consider different perspectives and weigh the potential consequences of our actions on others.
Making moral decisions: The PFC integrates emotional and cognitive information to guide our moral choices and actions.
This integration of empathy and moral reasoning allows us to act in ways that are consistent with our values and promote the well-being of others.

Examples of PFC Involvement in Empathy

Studies on individuals with damage to the PFC have provided valuable insights into its role in empathy. For example:

Phineas Gage: The famous case of Phineas Gage, a railroad worker who suffered severe damage to his PFC in an accident, revealed the importance of this brain region for social behavior and emotional regulation. After the accident, Gage exhibited impulsive behavior, poor social judgment, and a lack of empathy.

Lesion studies: Studies on individuals with lesions in different areas of the PFC have shown that damage to the ventromedial PFC can impair emotional regulation and social decision-making, while damage to the dorsolateral PFC can disrupt cognitive empathy and perspective-taking.

Mindfulness Practices and PFC Function

Mindfulness practices, such as meditation and yoga, have been shown to strengthen PFC function and enhance emotional regulation. By training our attention and cultivating self-awareness, mindfulness can help us to:

Increase PFC activity: Studies have shown that mindfulness meditation can increase activity in the PFC, particularly in areas involved in attention, emotional regulation, and self-awareness.

Improve emotional regulation: Mindfulness practices can help us to become more aware of our emotions and develop strategies to manage them effectively.

Enhance cognitive empathy: By cultivating a non-judgmental and compassionate mindset, mindfulness can enhance our ability to understand and share the feelings of others.

Recent Advances and Future Directions (2025)

Research on the PFC and its role in empathy continues to advance, with new findings in 2025 highlighting:

PFC connectivity: Studies are exploring how the PFC interacts with other brain regions, such as the limbic system and the DMN, to support empathy and social cognition.

Neuroplasticity of the PFC: Research is demonstrating the remarkable plasticity of the PFC, suggesting that its function can be enhanced through training and experience.

Neurofeedback and brain stimulation: Emerging technologies such as neurofeedback and brain stimulation are being investigated as potential tools for enhancing PFC function and improving empathy.
Expert Insights:

Dr. Antonio Damasio, a renowned neuroscientist, emphasizes the role of the PFC in integrating emotions and cognition to guide social behavior.
Dr. Richard Davidson, a leading researcher in affective neuroscience, highlights the importance of the PFC in emotional regulation and its potential to be strengthened through mindfulness practices.
Conclusion:

The PFC plays a pivotal role in empathy by regulating emotions, enabling cognitive empathy, and guiding moral judgment. By integrating emotional and cognitive information, the PFC allows us to navigate the complexities of social interactions and act with compassion and understanding. Further research on the PFC promises to deepen our understanding of empathy, social behavior, and the human mind.

Empathy in Child Development

2.1 Temperament and Early Empathic Tendencies: Inborn and Environmental Influences

The Interplay of Nature and Nurture

A child's capacity for empathy is shaped by a complex interplay of genetic and environmental factors. While some children may be predisposed towards empathy due to their innate temperament, early experiences and interactions with caregivers play a crucial role in nurturing and developing this innate potential.

Temperament and Empathy

Temperament refers to the individual behavioral and emotional characteristics that are present from early childhood. Certain temperamental traits have been linked to a child's capacity for empathy.

Sensitivity: Children who are highly sensitive to sensory stimuli and emotional cues may be more attuned to the feelings of others, making them more likely to respond with empathy.
Reactivity: Children who exhibit high emotional reactivity may be more easily overwhelmed by the distress of others, which could either lead to empathetic concern or withdrawal, depending on their coping mechanisms and the support they receive.
Sociability: Children who are naturally sociable and enjoy interacting with others are more likely to have opportunities to observe and understand different emotional expressions, promoting empathic development.

The Role of Parenting

Parents play a vital role in shaping a child's empathic tendencies. Sensitive and responsive parenting, where caregivers acknowledge and validate a child's emotions, provides a secure base for the child to explore their own feelings and understand the emotions of others. Parents who model empathy and compassion in their own interactions teach children the importance of caring for others.

Adapting Parenting Styles

Parents can adapt their parenting styles to support their child's unique temperamental needs and foster empathy:

For highly sensitive children: Parents can provide a calm and predictable environment, offer extra support during emotionally challenging situations, and help the child regulate their emotions effectively.
For highly reactive children: Parents can help the child develop coping strategies for managing intense emotions, teach them how to recognize and respond to the emotions of others in a safe and controlled manner.
For sociable children: Parents can encourage positive social interactions, provide opportunities for cooperative play, and guide the child in understanding different perspectives and resolving conflicts peacefully.

Longitudinal Studies

Longitudinal studies have provided valuable insights into the development of empathy in children with different temperaments. For instance, research has shown that children who are temperamentally fearful and inhibited may be more likely to develop empathy if they have supportive and nurturing caregivers.

Expert Insights

Experts emphasize the importance of early intervention in fostering empathy. Dr. Rebecca Saxe, a leading researcher in the field of cognitive neuroscience, highlights the plasticity of the developing brain and the potential for nurturing empathy through positive experiences and interventions.

Research-Backed Information

Research has consistently shown that empathy can be cultivated through interventions that focus on emotional literacy, perspective-taking, and prosocial skills. Programs like Roots of Empathy, which involve bringing babies into classrooms, have demonstrated significant improvements in children's empathy and prosocial behavior.

Examples and Case Studies

A case study of a child with a shy and inhibited temperament who developed strong empathic skills through participation in a drama program that encouraged role-playing and emotional expression.
An example of a parent who adapted their parenting style to support their highly reactive child by providing a calm and structured environment and teaching them relaxation techniques.

Conclusion

The development of empathy is a dynamic process that is influenced by both nature and nurture. By understanding the interplay of temperament and environment, parents and caregivers can create a supportive context that fosters empathy in children, laying the foundation for positive social and emotional development.

2.2 The Role of Play in Developing Empathy: Pretend Play, Social Play, and Perspective-Taking

Play as a Catalyst for Empathy

Play is not merely a recreational activity for children; it is a powerful tool for social and emotional development, particularly in fostering empathy. Through play, children learn to understand and share the feelings of others, negotiate different perspectives, and develop essential social skills.

Pretend Play and Perspective-Taking

Pretend play, also known as imaginative or symbolic play, involves children taking on different roles and scenarios, often imitating real-life situations or creating fantastical worlds. This type of play encourages children to step outside of themselves and imagine the thoughts and feelings of others. By embodying different characters, children learn to see the world from multiple perspectives, which is a crucial component of empathy.

Social Play and Cooperation

Social play involves interactions with other children, requiring cooperation, negotiation, and compromise. Through social play, children learn to navigate social dynamics, understand social cues, and respond appropriately to the emotions of others. They learn to share, take turns, and resolve conflicts, all of which contribute to the development of empathy and prosocial behavior.

Types of Play and Their Contributions to Empathy

Socio-dramatic play: This type of play involves children enacting social roles and scenarios, such as playing doctor and patient or teacher and student. It allows children to explore different social roles and perspectives, understand social norms, and practice empathy in a safe and controlled environment.
Constructive play: This involves building and creating things together, fostering collaboration, problem-solving, and shared understanding. Children learn to negotiate ideas, compromise, and work towards a common goal, promoting empathy and teamwork.
Games with rules: Playing games with rules teaches children about fairness, turn-taking, and following social norms. It also provides opportunities for children to understand and manage their own emotions, as well as recognize and respond to the emotions of others in the context of the game.

Observational Studies

Observational studies have provided compelling evidence for the link between play and empathy. Research has shown that children who engage in more pretend play tend to have better perspective-taking abilities and higher levels of empathy. Similarly, children who participate in cooperative social play are more likely to exhibit prosocial behaviors and show concern for others.

Facilitating Play Experiences

Parents and educators can facilitate play experiences that promote empathy by:

Providing a rich and stimulating play environment: Offer a variety of toys and materials that encourage different types of play, such as dolls, blocks, costumes, and art supplies.

Encouraging pretend play: Join in children's pretend play, taking on different roles and modeling empathy and perspective-taking.

Facilitating social play: Create opportunities for children to play together, providing guidance and support in navigating social interactions and resolving conflicts.

Talking about emotions: Help children identify and understand their own emotions and the emotions of others during play.

Expert Insights

Experts in child development emphasize the crucial role of play in fostering empathy. Dr. Stuart Brown, founder of the National Institute for Play, argues that play is essential for healthy social and emotional development, promoting creativity, problem-solving, and empathy.

Research-Backed Information

Research has consistently shown that play has a positive impact on children's social and emotional development. A study published in the journal Child Development found that children who engaged in more pretend play were better able to understand the perspectives of others and exhibited more prosocial behavior.

Examples and Case Studies

A case study of a child who struggled with social skills but showed significant improvement in empathy and social interaction after participating in a playgroup that focused on cooperative games and role-playing.

An example of a teacher who used puppets and storytelling to encourage pretend play and help children explore different emotions and perspectives.

Conclusion

Play is not just fun and games; it is a fundamental human activity that plays a vital role in the development of empathy. By providing opportunities for pretend play, social play, and perspective-taking, parents and educators can nurture children's capacity for empathy, laying the foundation for positive social relationships and emotional well-being.

2.3 Empathy and Prosocial Behavior in Children: Helping, Sharing, and Comforting

The Link Between Empathy and Prosocial Behavior

Empathy is not merely an emotional experience; it can motivate action. Prosocial behavior, which encompasses actions intended to benefit others, is closely linked to empathy. When children feel empathy for someone in need, they are more likely to engage in helping, sharing, and comforting behaviors.

Cognitive and Emotional Factors

Several cognitive and emotional factors contribute to children's willingness to engage in prosocial behavior:

Perspective-taking: The ability to understand another person's perspective and recognize their needs is essential for prosocial behavior. Children who can take another's perspective are more likely to recognize when someone needs help and offer assistance.
Emotional understanding: Children who can accurately identify and understand the emotions of others are better equipped to respond with empathy and offer appropriate support.

Moral reasoning: As children develop moral reasoning, they begin to understand concepts of fairness, justice, and helping those in need. This moral compass can guide their prosocial actions.
Self-efficacy: Children who believe in their ability to make a difference and help others are more likely to engage in prosocial behavior.
Influences on Helping Behavior

Several factors influence children's helping behavior:

Age: As children get older, their capacity for empathy and prosocial behavior increases. Older children are more likely to understand the needs of others and take action to help.
Emotional state: Children who are in a positive emotional state are more likely to help others. Conversely, children who are feeling sad or distressed may be less likely to engage in prosocial behavior.
Relationship to the person in need: Children are more likely to help people they know and care about, such as family members and friends. However, empathy can also extend to strangers, particularly when children witness someone in distress.
Encouraging Prosocial Behavior

Parents and educators can encourage prosocial behavior in children by:

Modeling prosocial behavior: Children learn by observing the adults in their lives. When parents and educators demonstrate kindness, compassion, and helpfulness, children are more likely to internalize these values and act accordingly.
Creating opportunities for helping: Provide children with opportunities to help others, both at home and in the

community. This could involve chores, volunteering, or simply assisting someone in need.
Praising prosocial behavior: When children engage in prosocial behavior, acknowledge and praise their actions. This positive reinforcement can encourage them to continue helping others.
Discussing prosocial values: Talk to children about the importance of kindness, empathy, and helping others. Encourage them to think about how their actions can impact others.

Experimental Studies

Experimental studies have shed light on the factors that influence children's helping behavior. For example, research has shown that children are more likely to help someone who is in distress if they believe they have the ability to make a difference.

Expert Insights

Experts in child development emphasize the importance of nurturing prosocial behavior from an early age. Dr. Nancy Eisenberg, a leading researcher in the field of prosocial development, highlights the role of empathy, perspective-taking, and moral reasoning in promoting helping behavior.

Research-Backed Information

Research has consistently shown that prosocial behavior is associated with positive outcomes for children, including increased social competence, academic achievement, and psychological well-being.

Examples and Case Studies

A case study of a child who overcame his shyness to help a classmate who was being bullied, demonstrating the power of empathy and moral courage.
An example of a school that implemented a program to promote prosocial behavior, resulting in increased helping, sharing, and cooperation among students.

Conclusion

Empathy and prosocial behavior are intertwined. By fostering empathy in children, we can encourage them to engage in actions that benefit others, creating a more compassionate and caring society. Parents and educators play a crucial role in nurturing prosocial development, providing children with the tools and opportunities to become kind, helpful, and contributing members of their communities.

2.4 The Impact of Adverse Childhood Experiences on Empathy: Trauma, Neglect, and Disrupted Development

Early Adversity and its Effects on Empathy

Adverse childhood experiences (ACEs) encompass a range of traumatic events that can have a profound impact on a child's development, including their capacity for empathy. These experiences can include physical, emotional, or sexual abuse, neglect, witnessing domestic violence, or growing up in a household with substance abuse or mental illness.

Disrupted Development

ACEs can disrupt the development of empathy in several ways:

Neurobiological effects: Trauma and neglect can alter brain development, particularly in regions associated with emotional regulation, stress response, and social cognition. These changes can impair a child's ability to understand and respond to the emotions of others.
Attachment disruptions: Secure attachment relationships with caregivers are essential for healthy emotional development, including empathy. ACEs can disrupt these attachments, leading to difficulties in forming trusting relationships and understanding the emotional needs of others.
Emotional dysregulation: Children who experience trauma often struggle with emotional regulation, making it challenging to manage their own emotions and respond appropriately to the emotions of others.
Social and cognitive deficits: ACEs can lead to social and cognitive deficits, hindering a child's ability to understand social cues, perspective-take, and engage in prosocial behavior.

Long-Term Effects of Childhood Trauma

Research has shown that childhood trauma can have long-term effects on brain development and social-emotional functioning. Individuals who have experienced ACEs are at increased risk for mental health problems, substance abuse, and difficulties in forming healthy relationships.

The Importance of Early Intervention

Early intervention is crucial for mitigating the negative effects of ACEs and promoting healthy development.

Interventions that focus on building resilience, fostering secure attachments, and providing trauma-informed care can help children who have experienced adversity to develop empathy and overcome the challenges associated with their experiences.

Interventions for Fostering Empathy and Resilience

Several interventions have shown promise in helping children who have experienced ACEs:

Trauma-focused therapy: This type of therapy helps children process traumatic experiences and develop coping strategies for managing their emotions.
Attachment-based interventions: These interventions focus on strengthening the bond between the child and their caregiver, promoting secure attachment and emotional regulation.
Play therapy: Play therapy provides a safe and supportive environment for children to express their emotions, process trauma, and develop social skills.
Mindfulness-based interventions: Mindfulness practices can help children develop self-awareness, emotional regulation, and empathy.
Research on the Effects of Childhood Trauma

Research has consistently shown a link between ACEs and a range of negative outcomes, including impaired empathy, increased risk for mental health problems, and difficulties in forming healthy relationships.

Expert Insights

Experts in child development and trauma emphasize the importance of early intervention and trauma-informed care for children who have experienced ACEs. Dr. Bruce Perry, a renowned expert in childhood trauma, highlights the need for interventions that address the neurobiological effects of trauma and promote resilience.

Examples and Case Studies

A case study of a child who experienced neglect and abuse and struggled with empathy and social skills but showed significant improvement after receiving trauma-focused therapy and participating in a mentoring program.
An example of a community program that provides support and resources for families affected by ACEs, promoting resilience and positive development in children.
Conclusion

Adverse childhood experiences can have a profound impact on a child's development, including their capacity for empathy. Early intervention and trauma-informed care are essential for mitigating the negative effects of ACEs and fostering resilience. By providing children with the support they need to heal and thrive, we can help them overcome the challenges associated with their experiences and develop into empathetic and compassionate individuals.

3.1 Empathy and Teacher-Student Relationships: Creating a Supportive and Inclusive Classroom Climate
The Power of Connection: Why Positive Teacher-Student Relationships Matter

In the tapestry of education, the threads of empathy weave a powerful bond between teachers and students, creating a vibrant and supportive classroom climate. At the heart of effective teaching lies the ability to understand and share the feelings of our students, fostering a sense of belonging and connection that fuels their learning journey.

Research consistently demonstrates the profound impact of positive teacher-student relationships on student engagement and academic achievement. When students feel seen, heard, and valued by their teachers, they are more likely to thrive academically, socially, and emotionally. A meta-analysis of 26 studies involving over 1,000 classrooms found that positive teacher-student relationships were significantly associated with increased student engagement, motivation, and academic performance (Roorda et al., 2011).

Building Bridges of Understanding: Strategies for Cultivating Positive Relationships

Teachers can cultivate positive and supportive relationships with their students by employing a range of strategies that prioritize empathy and understanding:

Active Listening: Truly listening to students, paying attention not only to their words but also to their nonverbal cues, conveys respect and fosters a sense of trust. By actively listening, teachers demonstrate that they value their students' thoughts, feelings, and perspectives.

Nonverbal Communication: A warm smile, a gentle nod, or eye contact can communicate volumes. Nonverbal cues can convey empathy, encouragement, and support, creating a positive and welcoming classroom environment.

Genuine Interest: Showing a genuine interest in students' lives beyond academics helps build rapport and strengthens the teacher-student connection. Taking the time to learn about their hobbies, interests, and families demonstrates that teachers care about them as individuals.

Empathetic Communication: Communicating with empathy involves understanding and acknowledging students' feelings, even when they are challenging or difficult. By validating their emotions and perspectives, teachers can create a safe and supportive space for students to express themselves.

High Expectations with Support: Communicating high expectations while providing the necessary support helps students feel challenged and capable. When students believe in their ability to succeed, they are more likely to persevere through challenges and achieve their goals.

Case Study: The Transformative Power of Empathy

Ms. Johnson, a fifth-grade teacher, noticed that one of her students, Sarah, was withdrawn and disengaged in class. Sarah's grades were slipping, and she seemed to be struggling both academically and socially. Ms. Johnson decided to reach out to Sarah, taking the time to listen to her concerns and understand her perspective. She discovered that Sarah was experiencing bullying from some of her classmates, which was affecting her self-esteem and her ability to focus on her studies.

Ms. Johnson responded with empathy and compassion, assuring Sarah that she was not alone and that she would help her address the situation. She worked with Sarah to develop strategies for coping with the bullying, and she also involved the school counselor to provide additional support. Ms. Johnson also spoke with the class about the importance of kindness and respect, emphasizing the impact that their words and actions can have on others.

Over time, Sarah began to feel more comfortable and confident in class. Her grades improved, and she started to participate more actively in group activities. Ms. Johnson's empathetic approach not only helped Sarah overcome a challenging situation but also created a more positive and inclusive classroom climate for all of her students.

Expert Insights:

Dr. Rita Pierson, a renowned educator and advocate for the importance of human connection in education, emphasizes the power of relationships in fostering student success. She states, "Kids don't learn from people they don't like." Building positive relationships with students is essential for creating a learning environment where they feel safe, supported, and motivated to learn.

Dr. Brene Brown, a research professor who has spent decades studying courage, vulnerability, and shame, highlights the importance of empathy in creating a sense of belonging. She states, "Empathy is not just something that happens to us - it's also a choice we make: to pay attention, to tune in to others." By choosing to empathize with our students, we create a classroom culture where everyone feels valued and respected.

Looking Ahead:

As we navigate the ever-evolving landscape of education, empathy remains a timeless and essential compass. By cultivating positive teacher-student relationships, we create a ripple effect that extends far beyond the classroom walls, shaping the lives of our students and empowering them to reach their full potential.

3.2 Empathy and Classroom Management: Moving Beyond Punishment to Understanding and Connection
Shifting the Paradigm: From Punishment to Understanding

Traditional classroom management often relies on punitive measures to control student behavior. However, this approach can create a climate of fear and resentment, hindering students' social-emotional development and academic progress. Empathy-based classroom management offers a powerful alternative, shifting the focus from punishment to understanding and connection. By seeking to understand the root causes of challenging behaviors, teachers can create a more positive and effective learning environment where students feel safe, respected, and supported.

Understanding the "Why" Behind Behavior:

Challenging behaviors are often a manifestation of unmet needs. Students may act out because they are struggling with academic difficulties, social-emotional challenges, or personal stressors. By taking the time to understand the "why" behind their behavior, teachers can address the underlying issues and provide the necessary support.

Empathy-Driven Strategies for Classroom Management:

Restorative Justice: This approach focuses on repairing harm and building relationships rather than simply punishing offenders. It involves bringing together those affected by a conflict to discuss the impact of their actions and collectively develop solutions.

Positive Behavior Interventions and Supports (PBIS): PBIS is a proactive framework that emphasizes teaching and reinforcing positive behaviors, creating a school-wide culture of respect and responsibility.

Conflict Resolution Strategies: Teaching students effective conflict resolution skills empowers them to resolve disagreements peacefully and respectfully. These strategies may include active listening, communication skills, and compromise.

Creating a Safe and Supportive Classroom Culture: Students need to feel safe expressing their emotions and needs without fear of judgment or punishment. Teachers can create this environment by fostering open communication, validating students' feelings, and providing opportunities for emotional expression.

Case Study: Transforming Conflict with Empathy

In Mr. Thompson's eighth-grade class, two students, David and Michael, were engaged in an ongoing conflict. Their disagreements often disrupted class and created a tense atmosphere. Instead of resorting to traditional disciplinary measures, Mr. Thompson decided to implement a restorative justice approach.

He facilitated a conversation between David and Michael, providing a safe space for them to express their perspectives and understand the impact of their actions on each other and the class. Through this process, the students were able to identify the root causes of their conflict and develop a plan for moving forward. They agreed to communicate more respectfully and to seek help from Mr. Thompson if conflicts arose in the future.

The restorative justice approach not only resolved the conflict between David and Michael but also created a more positive and collaborative classroom culture. Students learned the value of empathy, communication, and conflict resolution, skills that would serve them well beyond the classroom.

Expert Insights:

Dr. Ross Greene, a renowned clinical psychologist and author of "The Explosive Child," emphasizes the importance of collaborative problem-solving in addressing challenging behaviors. He states, "Kids do well if they can." When students struggle with their behavior, it is often because they lack the skills to meet expectations. By working collaboratively with students to identify and address the underlying challenges, teachers can help them develop the skills they need to succeed.

Dr. Daniel Siegel, a clinical professor of psychiatry and author of "Brainstorm: The Power and Purpose of the Teenage Brain," highlights the importance of understanding the adolescent brain in developing effective classroom management strategies. He explains that the adolescent brain is undergoing significant changes, which can impact their behavior and decision-making. By understanding these developmental changes, teachers can respond to challenging behaviors with empathy and compassion.

Looking Ahead:

As we continue to evolve our understanding of child development and learning, empathy-based classroom management offers a promising path forward. By moving beyond punishment and embracing understanding and connection, we can create learning environments where all students feel safe, respected, and empowered to reach their full potential.

3.3 Empathy and Inclusive Education: Meeting the Needs of Diverse Learners

Embracing Diversity: Creating a Welcoming and Supportive Learning Environment for All

In today's increasingly diverse classrooms, fostering an inclusive environment where every student feels valued and respected is paramount. Empathy plays a crucial role in creating such an environment, enabling teachers to understand and appreciate the unique needs, backgrounds, and experiences of their students. By embracing diversity and promoting equity, teachers can unlock the potential of every learner, regardless of their background or abilities.

Understanding the Needs of Diverse Learners:

Cultural Responsiveness: Culturally responsive teaching recognizes and values the cultural backgrounds of students, incorporating their experiences and perspectives into the curriculum. This approach fosters a sense of belonging and helps students connect with the learning material on a deeper level.

Students with Disabilities: Students with disabilities may require specific accommodations and support to access the curriculum and participate fully in the classroom. Empathetic teachers take the time to understand their individual needs and work collaboratively with them, their families, and support staff to create a learning environment that is accessible and inclusive.

Linguistic Diversity: In classrooms with students who are English language learners (ELLs), empathy is essential for creating a supportive and welcoming environment. Teachers can use a variety of strategies to support ELLs, such as providing visual aids, using clear and concise language, and offering opportunities for peer interaction.

Socioeconomic Diversity: Students from diverse socioeconomic backgrounds may face unique challenges that impact their learning. Empathetic teachers are sensitive to these challenges and strive to create an equitable learning environment where all students have access to the resources and support they need to succeed.

Strategies for Promoting Inclusive Education:

Universal Design for Learning (UDL): UDL is a framework that guides the development of flexible learning environments that can accommodate the diverse needs of all learners. It emphasizes providing multiple means of representation, engagement, and expression, allowing students to access and interact with the learning material in ways that work best for them.

Differentiated Instruction: Differentiated instruction involves tailoring instruction to meet the individual needs of students. 1 This may involve providing different learning activities, materials, or assessments based on students' readiness levels, interests, and learning profiles.

Building Relationships: Creating strong relationships with students is foundational to inclusive education. By getting to know their students as individuals, teachers can better understand their needs, interests, and strengths, allowing them to create a learning environment where everyone feels valued and respected.

Case Study: Empathy in Action: Supporting a Student with Autism

Ms. Rodriguez, a second-grade teacher, had a student named Alex who was diagnosed with autism spectrum disorder (ASD). Alex struggled with social interactions and often had difficulty focusing in class. Ms. Rodriguez was committed to creating an inclusive environment where Alex felt supported and successful.

She worked closely with Alex's parents and the school's special education team to develop an individualized education program (IEP) that addressed his specific needs. She incorporated visual supports into her lessons, provided sensory breaks throughout the day, and created a quiet corner in the classroom where Alex could go to decompress when feeling overwhelmed. Ms. Rodriguez also made an effort to pair Alex with a buddy during group activities, providing him with additional social support.

Through her empathetic approach and willingness to adapt her teaching practices, Ms. Rodriguez created a welcoming and supportive learning environment for Alex. He began to thrive in the classroom, making academic progress and developing positive relationships with his peers.

Expert Insights:

Dr. Temple Grandin, a renowned autism advocate and author of "The Autistic Brain," emphasizes the importance of understanding the sensory sensitivities of individuals with autism. She states, "The world needs all kinds of minds." By creating sensory-friendly classrooms, teachers can help students with autism feel more comfortable and focused, allowing them to learn and participate fully.

Dr. Carol Dweck, a Stanford University psychologist and author of "Mindset: The New Psychology of Success," highlights the importance of fostering a growth mindset in students. A growth mindset is the belief that abilities can be developed through hard work and dedication. By encouraging students to embrace challenges and view mistakes as opportunities for learning, teachers can create a more inclusive and supportive learning environment for all.

Looking Ahead:

As our classrooms become increasingly diverse, empathy will continue to be a guiding principle in creating inclusive and equitable learning environments. By embracing the unique needs of every learner and fostering a sense of belonging, we can empower all students to reach their full potential.

3.4 Empathy and Social-Emotional Learning (SEL): Integrating Empathy into the Curriculum

Empathy: The Cornerstone of Social-Emotional Learning

Social-emotional learning (SEL) has emerged as a critical component of education, equipping students with the essential skills to navigate the complexities of life. Empathy serves as the cornerstone of SEL, enabling students to understand and connect with others on a deeper level, fostering positive relationships, and contributing to a more compassionate and inclusive society.

The Role of Empathy in SEL Programs:

SEL programs typically focus on five core competencies: self-awareness, self-management, social awareness, relationship skills, and responsible decision-making. Empathy plays a pivotal role in each of these areas:

Self-awareness: By understanding their own emotions and perspectives, students can better understand the emotions and perspectives of others.

Self-management: Empathy helps students regulate their emotions and behaviors, enabling them to respond to challenging situations with compassion and understanding.

Social awareness: Empathy allows students to perceive and understand the emotions and perspectives of others, fostering a sense of connection and belonging.

Relationship skills: Empathy is essential for building and maintaining healthy relationships. It enables students to communicate effectively, resolve conflicts peacefully, and cooperate with others.

Responsible decision-making: Empathy helps students consider the impact of their actions on others, leading to more responsible and ethical choices.

Integrating Empathy Across the Curriculum:

Empathy can be woven into various subjects and activities across the curriculum, providing students with opportunities to develop and practice this essential skill.

Literature: Literature provides a powerful window into the human experience, allowing students to step into the shoes of characters and explore a range of emotions and perspectives. Through reading and discussing diverse stories, students can develop empathy and understanding for people from different backgrounds and experiences.

Art: Art provides a creative outlet for students to express their emotions and connect with others on an emotional level. Engaging in art activities, such as creating self-portraits, drawing pictures inspired by music, or sculpting emotions, can help students develop self-awareness and empathy.

Music: Music has the power to evoke strong emotions and create a sense of shared experience. Listening to and discussing music from different cultures and genres can help students develop empathy and appreciation for diverse perspectives.

History: Studying historical events and figures can provide valuable opportunities for students to develop empathy and understanding for people from different time periods and cultures. By exploring the motivations and perspectives of historical figures, students can gain a deeper understanding of the complexities of human behavior.

Science: Science can provide opportunities for students to explore the interconnectedness of life and develop empathy for the natural world. Studying ecosystems, animal behavior, or the impact of human actions on the environment can foster a sense of responsibility and stewardship.

Case Study: Cultivating Empathy Through Literature

Ms. Davis, a high school English teacher, incorporated a unit on empathy into her curriculum. She selected a variety of texts that explored themes of compassion, understanding, and overcoming adversity. Students read novels such as "To Kill a Mockingbird," "The Absolutely True Diary of a Part-Time Indian," and "The Hate U Give," engaging in discussions about the characters' experiences, perspectives, and motivations.

Ms. Davis also incorporated creative writing activities, asking students to write from the perspective of a character they felt empathy for or to create their own stories that explored themes of empathy and understanding. Through these activities, students deepened their understanding of empathy and its importance in building a more just and compassionate world.

Expert Insights:

Dr. Michele Borba, an educational psychologist and author of "UnSelfie: Why Empathetic Kids Succeed in Our All-About-Me World," emphasizes the importance of teaching empathy explicitly. She states, "Empathy is a skill, not a trait. It can be taught and learned." By providing students with opportunities to practice empathy in a variety of contexts, we can help them develop this essential skill.

Daniel Goleman, author of "Emotional Intelligence," highlights the connection between empathy and emotional intelligence. He states, "Empathy is the foundation of social awareness and relationship skills." By developing empathy, students can improve their ability to understand and connect with others, leading to more positive and fulfilling relationships.

Looking Ahead:

As we prepare students for the challenges and opportunities of the 21st century, empathy will continue to be a vital skill. By integrating empathy into the curriculum and providing students with opportunities to develop and practice this essential skill, we can empower them to become compassionate, responsible, and engaged citizens who contribute to a better world.

Empathy in Healthcare

4.1 Empathy and Patient Communication: Building Trust and Rapport

Empathy is a cornerstone of effective patient-provider communication. It involves understanding and sharing the feelings of another person, in this case, the patient. When healthcare providers demonstrate empathy, they create a safe and supportive environment where patients feel heard, understood, and respected. This, in turn, builds trust and rapport, which are essential for positive health outcomes.

Active Listening

Active listening is a critical component of empathetic communication. It involves fully concentrating on what the patient is saying, both verbally and nonverbally, and making a conscious effort to understand their perspective. This means paying attention not just to the words being spoken, but also to the patient's tone of voice, body language, and facial expressions.

Healthcare providers can demonstrate active listening by:

Maintaining eye contact
Leaning in slightly to show engagement
Nodding their head to acknowledge understanding
Using verbal cues like "uh-huh" or "I see"
Summarizing the patient's concerns to ensure accurate understanding
Asking clarifying questions to delve deeper into the patient's experience

Nonverbal Cues

Nonverbal cues play a significant role in communication, often conveying more meaning than words alone. Healthcare providers can use nonverbal cues to enhance their communication and demonstrate empathy. These cues include:

Eye contact: Maintaining appropriate eye contact conveys interest and attentiveness. However, it's impcrtant to be mindful of cultural differences, as some cultures may view direct eye contact as disrespectful.
Body language: Open and relaxed body language, such as uncrossed arms and leaning forward, signals approachability and engagement. Mirroring the patient's posture can also create a sense of connection.
Facial expressions: Facial expressions should reflect empathy and understanding. Smiling, nodding, and showing concern through facial expressions can help patients feel more comfortable and at ease.

Clear and Compassionate Language

The language healthcare providers use can significantly impact patient understanding and emotional well-being. Clear and compassionate language involves:

Using plain language that is easy for patients to understand, avoiding medical jargon
Explaining medical concepts in a way that is accessible and relevant to the patient's situation
Speaking with a calm and reassuring tone of voice
Expressing empathy and concern through words, such as "I understand this must be difficult for you" or "I'm here to support you through this"

Empathetic Communication Techniques

There are several specific communication techniques that healthcare providers can use to demonstrate empathy:

Reflecting: This involves mirroring back the patient's feelings to show understanding. For example, a provider might say, "You seem to be feeling anxious about this procedure."
Validating: This involves acknowledging the patient's feelings and experiences as legitimate. For example, a provider might say, "It's understandable that you're feeling frustrated."
Using open-ended questions: Open-ended questions encourage patients to share more information and express their thoughts and feelings. For example, instead of asking, "Are you in pain?" a provider might ask, "Can you tell me more about how you're feeling?"

Cultural Sensitivity and Health Literacy

Cultural sensitivity and awareness of health literacy levels are crucial for empathetic communication in diverse healthcare settings. Healthcare providers should be aware of:

Cultural beliefs and practices: Different cultures may have different beliefs about health and illness, as well as different communication styles. Providers should be respectful of these differences and adapt their communication accordingly.
Health literacy: Health literacy refers to a patient's ability to understand and use health information. Providers should assess patients' health literacy levels and adjust their communication to ensure understanding. This may involve using plain language, visual aids, or other communication tools.

Examples of Empathetic Communication

Scenario: A patient is diagnosed with a serious illness and is feeling overwhelmed and scared.

Empathetic response: "I can see that you're feeling overwhelmed right now. This is a lot to take in. I want you to know that I'm here for you and will support you every step of the way. We'll work together to develop a treatment plan that meets your needs and goals."

Scenario: A patient is hesitant to undergo a recommended procedure due to fear of complications.

Empathetic response: "I understand your concerns about the procedure. It's normal to feel anxious when facing a medical decision. Let's discuss the risks and benefits in more detail so you can make an informed decision that feels right for you."

Scenario: A patient is struggling to manage a chronic condition and is feeling frustrated and discouraged.

Empathetic response: "It sounds like you're going through a tough time right now. Managing a chronic condition can be challenging. I want you to know that you're not alone in this. We'll work together to find ways to support you and improve your quality of life."

Conclusion

Empathy is a fundamental skill for healthcare providers. By demonstrating empathy through active listening, nonverbal cues, and clear and compassionate language, providers can build trust with patients, reduce anxiety, and improve understanding. This, in turn, leads to better patient outcomes and a more positive healthcare experience.

4.2 Empathy and Medical Decision-Making: Navigating Ethical Dilemmas

Empathy plays a crucial role in ethical decision-making in healthcare, particularly in complex situations with significant implications for patients' lives. These situations often involve weighing competing values, considering potential risks and benefits, and making difficult choices that respect patient autonomy and well-being. By cultivating empathy, healthcare providers can gain a deeper understanding of patients' perspectives, values, and preferences, leading to more patient-centered and ethically sound decisions.

End-of-Life Care

End-of-life care often presents challenging ethical dilemmas, such as deciding whether to withdraw life support, initiate palliative care, or respect a patient's advance directives. These decisions can be emotionally charged and involve complex medical considerations. Empathy allows healthcare providers to connect with patients and their families on a human level, understand their fears and hopes, and make decisions that align with their values and wishes.

Informed Consent

Informed consent is a cornerstone of ethical healthcare practice. It requires that patients be fully informed about their medical condition, treatment options, potential risks and benefits, and alternative approaches. Empathy helps healthcare providers communicate this information in a way that is understandable and meaningful to patients, taking into account their individual needs and concerns. By fostering a trusting relationship and actively listening to patients' questions and concerns, providers can ensure that patients are truly empowered to make informed decisions about their care.

Treatment Options with Significant Risks and Benefits

Many medical treatments involve a balance of potential risks and benefits. For example, a cancer patient may need to weigh the potential benefits of chemotherapy against the risk of debilitating side effects. Empathy allows healthcare providers to understand the patient's individual circumstances, values, and preferences, and to tailor treatment recommendations accordingly. By engaging in shared decision-making and considering the patient's quality of life, providers can help patients make choices that are aligned with their overall goals and values.

Case Studies

Case Study 1: Withdrawing Life Support

A patient with a terminal illness is on life support. The medical team believes that further treatment is futile, but the patient's family is reluctant to withdraw life support. In this situation, empathy can help the healthcare team understand the family's emotional distress and their hope for a miracle. By acknowledging their grief and providing compassionate support, the team can facilitate a discussion about the patient's wishes and the goals of care. This may involve reviewing advance directives, exploring palliative care options, and offering spiritual or emotional support.

Case Study 2: Disclosing a Difficult Diagnosis

A patient receives a diagnosis of a serious illness with a poor prognosis. The healthcare provider must decide how to disclose this information in a way that is both truthful and compassionate. Empathy can guide the provider in delivering the news with sensitivity, acknowledging the patient's emotional response, and offering support and hope. This may involve providing information about treatment options, connecting the patient with support groups, and addressing their fears and concerns.

Case Study 3: Balancing Treatment Risks and Benefits

A patient with a chronic condition is faced with a treatment decision that involves significant risks and potential benefits. The healthcare provider must help the patient weigh these factors and make an informed decision. Empathy allows the provider to understand the patient's individual circumstances, values, and preferences. By engaging in shared decision-making and considering the patient's quality of life, the provider can help the patient make a choice that is aligned with their overall goals and values.

Conclusion

Empathy is a critical skill for healthcare providers navigating ethical dilemmas. By understanding and sharing patients' feelings, values, and perspectives, providers can make more informed and compassionate decisions that prioritize patient well-being and respect their autonomy. Empathy fosters trust, enhances communication, and promotes ethical practice in healthcare.

4.3 Empathy and Healthcare Teams: Fostering Collaboration and Communication

Empathy is not only essential in patient-provider interactions but also plays a vital role in fostering collaboration and communication within healthcare teams. In the complex and dynamic environment of healthcare, effective teamwork is crucial for providing safe and high-quality patient care. Empathy enables team members to understand each other's roles, perspectives, and challenges, leading to improved communication, mutual respect, and a shared sense of purpose.

Understanding Roles and Perspectives

Healthcare teams are composed of professionals from various disciplines, each with their own specialized knowledge, skills, and responsibilities. Empathy allows team members to appreciate the contributions of others and recognize the value of diverse perspectives. By understanding the challenges and pressures faced by colleagues, team members can offer support, collaborate effectively, and work towards shared goals.

Improved Communication

Empathy enhances communication within healthcare teams by promoting active listening, open dialogue, and constructive feedback. When team members feel understood and respected, they are more likely to communicate openly and honestly, share information freely, and address conflicts constructively. This leads to improved coordination of care, reduced errors, and better patient outcomes.

Mutual Respect and Trust

Empathy fosters a culture of mutual respect and trust within healthcare teams. When team members feel valued and appreciated, they are more likely to trust each other's judgment, support each other's decisions, and work together effectively. This creates a positive and supportive work environment, which can reduce stress, improve job satisfaction, and enhance team performance.

Team-Building Activities

Various team-building activities can promote empathy and communication within healthcare teams:

Simulations: Simulated scenarios allow team members to practice working together in a safe and controlled environment. This can help them develop communication skills, learn to anticipate each other's needs, and build trust in each other's abilities.
Role-playing exercises: Role-playing allows team members to step into the shoes of others and experience different perspectives. This can increase empathy and understanding, particularly in situations where there may be conflicting viewpoints or power dynamics.
Reflective practice sessions: Reflective practice encourages team members to reflect on their experiences, identify areas for improvement, and learn from each other. This can promote self-awareness, enhance communication skills, and foster a culture of continuous learning.
Interprofessional Education

Interprofessional education (IPE) brings together students from different healthcare disciplines to learn and work together. IPE fosters empathy and collaboration by providing opportunities for students to:

Learn about each other's roles and responsibilities: This helps students appreciate the contributions of different professions and understand how their roles intersect in patient care.
Develop communication and teamwork skills: IPE provides opportunities for students to practice communicating

effectively, resolving conflicts, and working collaboratively in a team setting.
Build respect for diverse perspectives: By interacting with students from other disciplines, students gain a broader understanding of healthcare and learn to value different viewpoints.
Examples of Empathy in Healthcare Teams

Scenario: A nurse notices that a physician seems stressed and overwhelmed. The nurse offers to assist with some tasks and provides a listening ear. This empathetic gesture helps to alleviate the physician's stress and fosters a supportive team environment.

Scenario: During a team meeting, a pharmacist raises concerns about a potential medication interaction. The team listens attentively, discusses the issue collaboratively, and arrives at a solution that ensures patient safety. This demonstrates effective communication and mutual respect within the team.

Scenario: A team of healthcare professionals is caring for a patient with complex needs. By sharing information openly, communicating effectively, and supporting each other's roles, the team provides coordinated and patient-centered care.

Conclusion

Empathy is essential for building strong and effective healthcare teams. By fostering understanding, communication, and mutual respect, empathy enables team members to work together collaboratively, provide safe and high-quality patient care, and create a positive and supportive work environment.

4.4 Empathy and Burnout Prevention: Protecting the Well-being of Healthcare Professionals

Healthcare professionals face numerous challenges in their daily work, including long hours, high workloads, exposure to suffering and death, and the emotional toll of caring for others. These factors can contribute to burnout, compassion fatigue, and secondary traumatic stress, which can have a significant impact on the well-being of healthcare providers and the quality of care they provide. Cultivating empathy, both for patients and for oneself, can play a crucial role in protecting healthcare professionals from these risks.

Empathy and Burnout

While empathy is essential for providing compassionate care, it can also be a source of stress and emotional exhaustion for healthcare providers. Witnessing the suffering of others, particularly in situations where there are limited treatment options or poor outcomes, can take a toll on providers' emotional well-being. This is especially true for those who are highly empathetic and deeply invested in their patients' well-being.

However, empathy can also be a protective factor against burnout. By understanding and sharing the emotions of others, healthcare providers can develop a sense of shared humanity and connection. This can help them to maintain perspective, avoid emotional detachment, and find meaning in their work.

Compassion Fatigue and Secondary Traumatic Stress

Compassion fatigue is a state of emotional and physical exhaustion that can occur in individuals who are exposed to the suffering of others. It is characterized by feelings of cynicism, detachment, and a reduced capacity for empathy. Secondary traumatic stress is a similar condition that can occur in individuals who are indirectly exposed to trauma, such as through witnessing or hearing about traumatic events.

Empathy can help healthcare providers to recognize the signs of compassion fatigue and secondary traumatic stress in themselves and others. By acknowledging their own emotional responses and seeking support when needed, providers can prevent these conditions from escalating and impacting their well-being and their ability to provide care.

Self-Awareness and Self-Care

Self-awareness is crucial for healthcare professionals to recognize their own emotional responses and identify potential triggers for burnout, compassion fatigue, and secondary traumatic stress. This involves paying attention to their own thoughts, feelings, and behaviors, and recognizing when they are feeling overwhelmed or emotionally depleted.

Self-care is essential for maintaining emotional and physical well-being. This includes prioritizing activities that promote relaxation, stress reduction, and emotional renewal. Examples of self-care practices include:

Mindfulness practices: Mindfulness involves paying attention to the present moment without judgment. This can help to reduce stress, increase self-awareness, and enhance emotional regulation.
Physical activity: Regular exercise can help to reduce stress, improve mood, and boost energy levels.

Healthy eating: Eating a balanced diet can provide the nutrients and energy needed to cope with stress and maintain well-being.
Adequate sleep: Getting enough sleep is essential for physical and emotional health.
Social support: Connecting with friends, family, and colleagues can provide emotional support and a sense of belonging.
Hobbies and leisure activities: Engaging in activities that bring joy and relaxation can help to recharge and restore emotional well-being.
Supportive Work Environments

Organizational factors can also play a significant role in preventing burnout and promoting the well-being of healthcare professionals. Supportive work environments are characterized by:

Reasonable workloads: Healthcare providers should have manageable workloads that allow them to provide quality care without feeling overwhelmed.
Adequate staffing levels: Sufficient staffing levels can ensure that providers have the support they need to manage their workload and provide safe patient care.
Opportunities for professional development: Providing opportunities for continuing education, skill development, and career advancement can enhance job satisfaction and reduce burnout.
Culture of compassion: A culture of compassion within the organization can foster a sense of shared purpose, support, and resilience among healthcare providers.
Access to mental health resources: Organizations should provide access to mental health services, such as counseling and support groups, to help providers cope with stress and emotional challenges.

Examples of Empathy in Burnout Prevention

Scenario: A nurse is feeling overwhelmed by the emotional demands of caring for terminally ill patients. The nurse attends a support group for healthcare professionals and finds solace in sharing experiences with colleagues who understand the challenges of the job.

Scenario: A physician is experiencing compassion fatigue and is struggling to connect with patients. The physician takes a mindfulness course and learns techniques for managing stress and cultivating self-compassion.

Scenario: A hospital implements a program to promote work-life balance for its employees. This includes offering flexible work schedules, on-site childcare, and access to wellness programs. These initiatives help to reduce stress and improve the well-being of healthcare providers.

Conclusion

Empathy plays a multifaceted role in preventing burnout and protecting the well-being of healthcare professionals. By cultivating empathy for patients, for themselves, and for their colleagues, healthcare providers can build resilience, manage stress, and find meaning in their work. Supportive work environments and access to mental health resources are also crucial for promoting the well-being of healthcare professionals and ensuring that they can continue tc provide compassionate and high-quality care.

Empathy in the Workplace

5.1 Empathy and Leadership Styles: Inspiring and Motivating Teams

Empathy is a critical skill for leaders who want to create positive and productive work environments. Empathetic leaders are able to understand and share the feelings of their team members, which allows them to build trust, foster collaboration, and empower their teams.

Different leadership styles incorporate empathy in different ways.

Transformational leaders inspire and motivate their teams by creating a shared vision and empowering them to achieve it. Empathy is essential for transformational leaders because it allows them to connect with their team members on a personal level and understand their needs and aspirations.
Servant leaders prioritize the needs of their team members and help them to develop and grow. Empathy is also essential for servant leaders because it allows them to understand the challenges their team members are facing and provide them with the support they need to succeed.
Authentic leaders are true to themselves and their values. Empathy is important for authentic leaders because it allows them to connect with their team members on a genuine level and build trust.
Examples of leaders who demonstrate empathy in their interactions with employees:

Indra Nooyi, former CEO of PepsiCo, is known for her empathetic leadership style. She regularly met with employees at all levels of the organization to get their feedback and ideas. She also made a point of recognizing individual contributions and celebrating team successes.

Satya Nadella, CEO of Microsoft, has also been praised for his empathetic leadership. He has focused on creating a more inclusive culture at Microsoft and has been vocal about the importance of empathy in the workplace.

Research-backed information on the benefits of empathetic leadership:

A study by the Center for Creative Leadership found that empathetic leaders are more likely to be seen as effective by their team members.

Another study by the University of Michigan found that empathetic leaders are more likely to create a positive work environment.

Experts' insights on the importance of empathy in leadership:

"Empathy is the most important leadership skill," says Simon Sinek, author of the book Leaders Eat Last. "It's the ability to understand and share the feelings of others. When you can do that, you can connect with people on a deeper level and build trust."

"Empathy is not just a nice-to-have skill for leaders," says Brené Brown, author of the book Daring Greatly. "It's a must-have skill. It's the key to creating a culture of trust and belonging."

5.2 Empathy and Organizational Communication: Creating a Culture of Understanding

Empathy can also improve communication within organizations. When people feel understood, they are more likely to communicate openly and honestly. This can lead to greater clarity, understanding, and collaboration.

Empathy can help to bridge communication gaps in a number of ways.

It can help people to understand different perspectives.
It can help people to communicate more effectively.
It can help to resolve conflict.
Specific strategies for using empathy in workplace communication:

Active listening: This involves paying attention to what the other person is saying, both verbally and nonverbally.
Perspective-taking: This involves trying to see the situation from the other person's point of view.
Nonviolent communication: This is a communication style that focuses on expressing needs and feelings in a way that is respectful and compassionate.
Organizations can foster a culture of empathy through a number of initiatives.

Communication training: This can help employees to develop their empathy skills.
Feedback mechanisms: This can help employees to understand how their communication is being received by others.
Leadership development programs: This can help leaders to develop their empathy skills and create a more empathetic workplace culture.
Examples of organizations that have successfully implemented empathy-building initiatives:

Google has a number of programs in place to help employees develop their empathy skills. These include workshops on active listening and perspective-taking.
The Ford Motor Company has implemented a program called "Empathy Training for Leaders." This program helps leaders to understand the importance of empathy and

how to use it to improve their communication and leadership skills.
Research-backed information on the benefits of empathy in organizational communication:

A study by the Harvard Business Review found that companies with high levels of empathy are more likely to have engaged employees.
Another study by the University of Pennsylvania found that empathy can help to improve team performance.
Experts' insights on the importance of empathy in organizational communication:

"Empathy is the key to effective communication," says Kim Scott, author of the book Radical Candor. "When you can understand and share the feelings of others, you can communicate in a way that is clear, concise, and compassionate."
"Empathy is the foundation of a strong workplace culture," says Adam Grant, author of the book Give and Take. "When people feel understood and respected, they are more likely to be engaged and productive."

5.3 Empathy and Negotiation: Reaching Mutually Beneficial Agreements

Negotiation is a process of communication between two or more parties who are trying to reach an agreement. Empathy can be a valuable tool in negotiation because it helps parties to understand each other's needs, interests, and perspectives. This understanding can facilitate compromise, build trust, and lead to mutually beneficial agreements.

How Empathy Improves Negotiation Outcomes

Understanding Needs and Interests: By understanding the other party's underlying needs and interests, you can identify common ground and potential areas for compromise. This is often more effective than focusing solely on stated positions.
Building Trust and Rapport: Empathy helps to build trust and rapport between parties. When people feel understood, they are more likely to be open to collaboration and compromise.
Reducing Conflict: Empathy can help to de-escalate conflict by allowing parties to see the situation from each other's perspective. This can help to reduce defensiveness and promote understanding.
Finding Creative Solutions: By understanding each other's needs, empathetic negotiators can often find creative solutions that meet the interests of all parties involved.
Negotiation Strategies that Prioritize Empathy

Active Listening: Pay close attention to what the other party is saying, both verbally and nonverbally. Ask clarifying questions to ensure you understand their perspective.
Reframing: Restate the other party's concerns in your own words to demonstrate understanding and show that you are actively listening.
Finding Common Ground: Identify shared goals and interests to build a foundation for collaboration.
Perspective-Taking: Consciously try to see the situation from the other party's point of view. This can help you understand their motivations and concerns.
Emotional Intelligence: Be aware of your own emotions and the emotions of the other party. Manage your emotions effectively and respond to the other party's emotions with empathy.

Examples of Empathy in Negotiation

Labor Dispute: In a labor dispute, a union representative who understands the financial pressures facing the company may be more willing to compromise on certain demands. Similarly, a company representative who empathizes with the workers' need for fair wages and benefits is more likely to find a mutually agreeable solution.
Business Deal: When negotiating a business deal, a salesperson who takes the time to understand the customer's needs and challenges is more likely to close the deal. By demonstrating empathy and offering solutions that address the customer's specific concerns, the salesperson builds trust and creates a win-win situation.
Case Study: The Camp David Accords

The Camp David Accords, a peace treaty between Israel and Egypt signed in 1978, is a prime example of empathy in negotiation. President Jimmy Carter facilitated the negotiations by creating a space for both Menachem Begin and Anwar Sadat to express their concerns and fears. Carter actively listened to both leaders, built rapport, and helped them find common ground, ultimately leading to a historic peace agreement.

Experts' Insights on Empathy in Negotiation

Chris Voss, former FBI hostage negotiator and author of Never Split the Difference, emphasizes the importance of tactical empathy in negotiation. He suggests using phrases like "It seems like..." or "It sounds like..." to demonstrate understanding and build rapport.
William Ury, co-author of Getting to Yes, highlights the importance of understanding the other party's BATNA (Best Alternative to a Negotiated Agreement) to effectively negotiate. This requires empathy to understand their options and motivations.

5.4 Empathy and Diversity & Inclusion: Creating an Equitable and Welcoming Workplace

Empathy is fundamental to fostering a truly diverse and inclusive workplace where all employees feel valued, respected, and supported. It allows individuals to connect with colleagues from different backgrounds, understand their experiences, and challenge their own biases. By promoting empathy, organizations can create an environment where everyone feels a sense of belonging and can contribute their best work.

How Empathy Supports Diversity & Inclusion

Breaking Down Stereotypes: Empathy encourages individuals to look beyond stereotypes and assumptions by understanding the unique experiences and perspectives of others. This helps to dismantle biases and create a more accepting environment.

Reducing Bias: By understanding the impact of bias and discrimination on individuals, empathy can help to reduce prejudice and promote fair treatment. It encourages individuals to challenge their own biases and advocate for equality.

Promoting Understanding: Empathy facilitates understanding across differences in race, ethnicity, gender, sexual orientation, religion, ability, and other dimensions of diversity. It fosters open dialogue and creates opportunities for learning and growth.

Building Trust and Respect: When employees feel understood and respected, they are more likely to trust their colleagues and leaders. This creates a more positive and productive work environment where everyone feels safe to be themselves.

Enhancing Collaboration: Empathy enables individuals to connect with colleagues on a deeper level, fostering collaboration and teamwork. When people feel

understood, they are more likely to work together effectively and achieve shared goals.
Diversity & Inclusion Initiatives that Leverage Empathy

Employee Resource Groups (ERGs): ERGs provide a platform for employees from diverse backgrounds to connect, share experiences, and support one another. They can also serve as valuable resources for the organization, offering insights and perspectives on diversity and inclusion initiatives.
Mentorship Programs: Mentorship programs can pair employees from different backgrounds, providing opportunities for learning and growth. Mentors can offer guidance and support, while mentees can gain valuable insights and perspectives.
Unconscious Bias Training: Unconscious bias training helps individuals to identify and challenge their own biases. By understanding how unconscious biases can impact behavior and decision-making, individuals can take steps to mitigate their effects and promote fairness.
Inclusive Leadership Development: Training leaders on inclusive leadership practices helps to create a culture of empathy and belonging. Leaders who are skilled in empathy can create a more welcoming and supportive environment for all employees.
Diversity & Inclusion Councils: These councils, composed of employees from different backgrounds, can provide guidance and feedback on diversity and inclusion initiatives. They can also help to ensure that the organization is meeting the needs of all employees.
Creating a Culture of Empathy and Inclusion

Leadership Commitment: Leaders must demonstrate a genuine commitment to diversity and inclusion. This includes setting clear expectations, providing resources, and holding individuals accountable for their behavior.

Inclusive Policies and Practices: Organizations should implement policies and practices that promote diversity and inclusion. This includes fair hiring practices, equitable compensation, and opportunities for advancement for all employees.
Open Communication: Create a culture of open communication where employees feel comfortable sharing their experiences and perspectives. This can be achieved through regular feedback mechanisms, employee surveys, and open forums.
Continuous Learning: Provide ongoing opportunities for employees to learn about diversity and inclusion. This can include workshops, training sessions, and online resources.
Examples of Organizations Promoting Empathy and Inclusion

Accenture: Accenture has a strong commitment to diversity and inclusion, with a focus on creating a culture of empathy and belonging. The company has a number of initiatives in place, including ERGs, mentorship programs, and unconscious bias training.
Salesforce: Salesforce is another company that is leading the way in diversity and inclusion. The company has a Chief Equality Officer and a dedicated team focused on creating a more inclusive workplace. Salesforce also has a number of programs in place to support employees from underrepresented groups.
Experts' Insights on Empathy, Diversity & Inclusion

Verna Myers, diversity and inclusion expert, emphasizes the importance of moving beyond simply tolerating diversity to actively celebrating it. This requires empathy to understand and appreciate the value of different perspectives and experiences.
Ruchika Tulshyan, author of The Diversity Advantage, *highlights the importance of inclusive leadership in* creating a workplace where everyone feels valued and

respected. Inclusive leaders demonstrate empathy, actively listen to their employees, and create opportunities for everyone to contribute.

By fostering a culture of empathy, organizations can create a more diverse, equitable, and inclusive workplace where everyone feels a sense of belonging and can thrive.

Empathy and Social Justice

6.1 Empathy and Moral Development: The Foundation of Ethical Action

Empathy, the ability to understand and share the feelings of others, plays a crucial role in moral development. It allows us to step outside our own perspectives and recognize the impact of our actions on others, fostering a sense of responsibility and ethical behavior. This section will delve into the relationship between empathy and moral development, exploring how empathy informs our understanding of right and wrong and motivates us to act in accordance with our values.

Kohlberg's Stages of Moral Development and the Role of Empathy

Lawrence Kohlberg, a renowned psychologist, proposed a stage theory of moral development that outlines how individuals progress through different levels of moral reasoning. His theory highlights the significance of empathy in advancing to higher levels of moral development. Let's examine Kohlberg's stages and the role of empathy in each:

Level 1: Preconventional Morality: At this level, moral reasoning is primarily driven by self-interest and avoiding punishment. Children in this stage may exhibit empathy, but it is often limited to their immediate circle and based on concrete consequences.
Level 2: Conventional Morality: In this stage, individuals conform to social norms and expectations to maintain order and gain approval. Empathy plays a more significant role as individuals begin to consider the perspectives and feelings of others within their social groups.

Level 3: Postconventional Morality: At this highest level individuals develop their own ethical principles based on universal human rights and justice. Empathy is crucial in this stage as it enables individuals to transcend their own cultural and social biases and recognize the inherent worth and dignity of all people.

Examples of Moral Courage and Empathy in Challenging Situations

History is replete with individuals who have demonstrated moral courage and empathy in challenging situations, standing up to injustice and advocating for marginalized groups. These individuals often act in accordance with their values, even when faced with opposition or personal risk. Here are a few examples:

Martin Luther King Jr.: A prominent leader of the Civil Rights Movement, King advocated for racial equality and justice through nonviolent resistance. His deep empathy for the suffering of African Americans fueled his unwavering commitment to social change.

Malala Yousafzai: A Pakistani activist for female education, Malala survived an assassination attempt by the Taliban for speaking out against their ban on girls' education. Her courage and empathy for girls denied access to education have inspired millions worldwide.

Nelson Mandela: An anti-apartheid revolutionary and former president of South Africa, Mandela spent 27 years in prison for his activism. His unwavering commitment to justice and reconciliation, even towards his former oppressors, exemplifies the power of empathy to overcome hatred and division.

The Importance of Empathy in Ethical Decision-Making

Empathy is not merely a passive emotion; it can actively inform our understanding of right and wrong and motivate us to act ethically. By considering the perspectives and feelings of others, we can make more informed and compassionate decisions that align with our values. Empathy can also help us to recognize and challenge our own biases and prejudices, leading to more equitable and just outcomes.

Conclusion

Empathy is an essential component of moral development, enabling us to transcend self-interest and recognize the inherent worth and dignity of all people. By fostering empathy, we can cultivate a more compassionate and just society where individuals are motivated to act in accordance with their values and stand up to injustice.

6.2 Empathy and Social Inequality: Understanding and Addressing Systemic Injustice

Social inequality, manifested in various forms like poverty, discrimination, and systemic oppression, remains a persistent challenge in societies worldwide. Empathy plays a critical role in raising awareness of these inequalities, motivating individuals to take action, and working towards a more equitable society. This section explores how empathy can be a powerful tool for understanding and addressing systemic injustice.

Empathy as a Catalyst for Recognizing Social Inequality

Often, individuals who are privileged or belong to dominant groups may remain unaware of the challenges faced by marginalized communities. Empathy acts as a bridge, allowing individuals to step outside their own experiences and understand the lived realities of others. By cultivating empathy, we can:

Develop awareness of systemic issues: Empathy helps us to recognize that social inequalities are not merely individual misfortunes but often stem from deeply entrenched systems of oppression and discrimination.
Challenge personal biases: We all carry unconscious biases that can perpetuate inequality. Empathy allows us to confront these biases and actively work towards dismantling them.
Humanize marginalized groups: Media portrayals and stereotypes often dehumanize marginalized communities. Empathy enables us to see beyond these stereotypes and recognize the shared humanity of all people.
Empathy as a Motivator for Social Justice Action

While awareness is crucial, empathy also serves as a powerful motivator for taking action against social injustice. When we genuinely connect with the struggles of others, we are more likely to:

Feel a sense of responsibility: Recognizing the suffering caused by inequality can evoke a sense of moral obligation to act.
Overcome apathy and inaction: Empathy can counteract feelings of helplessness and inspire individuals to become agents of change.
Engage in collective action: Empathy can foster solidarity and encourage individuals to join social justice movements and work towards systemic change.
Examples of Social Justice Movements Fueled by Empathy

Throughout history, numerous social justice movements have been driven by empathy and the desire to create a more just and equitable society. Some notable examples include:

The Civil Rights Movement: The struggle for racial equality in the United States was fueled by empathy for the

injustices faced by African Americans. Leaders like Martin Luther King Jr. effectively used empathy to mobilize support and bring about significant legal and social change.
The Rights Movement: The fight for rights has been propelled by empathy for the discrimination and marginalization faced by this community. Activists have successfully used empathy to challenge societal norms and achieve greater acceptance and equality.
The #MeToo Movement: The #MeToo movement, which brought widespread attention to sexual harassment and assault, was fueled by empathy for survivors and a collective desire to hold perpetrators accountable and create safer spaces for women.
Empathy as a Bridge Between Social Groups

Empathy can play a crucial role in bridging divides between different social groups and promoting understanding and solidarity. By fostering empathy, we can:

Challenge stereotypes and prejudices: Empathy allows us to see beyond simplistic generalizations and recognize the diversity and complexity of human experiences within different groups.
Promote dialogue and understanding: Empathy encourages open communication and active listening, fostering mutual respect and understanding between different groups.
Build coalitions for social change: Empathy can help to create a sense of shared purpose and encourage collaboration between diverse groups working towards common goals.
Conclusion

Empathy is not merely a passive emotion but a powerful tool for understanding and addressing social inequality. By cultivating empathy, we can develop awareness of systemic injustices, challenge our biases, and motivate ourselves to take action towards creating a more just and equitable society. Empathy can also bridge divides between different social groups, fostering understanding, solidarity, and collective action for social change.

6.3 Empathy and Activism: Fueling Social Change

Empathy plays a pivotal role in inspiring and sustaining activism, providing the fuel for individuals and groups to challenge injustice and strive for a better world. Activists leverage empathy to connect with others, mobilize support, and create meaningful social change. This section investigates how empathy functions as a driving force in activism and explores how activists can effectively utilize empathy to achieve their goals.

Empathy as an Inspiration for Activism

At the heart of activism lies a deep sense of empathy for those who are suffering or experiencing injustice. This empathy can manifest in various ways:

Personal connection: Activists may be personally affected by an issue or have close ties to individuals or communities experiencing hardship. This personal connection fuels their passion for change.
Witnessing injustice: Observing or learning about injustice can evoke strong empathetic responses, prompting individuals to take action to address the issue.
Moral outrage: Empathy can spark moral outrage at the unfairness or cruelty of a situation, motivating individuals to fight for justice and equality.

Empathy as a Tool for Mobilization

Activists utilize empathy to connect with and mobilize others, building support for their cause and creating a collective force for change. Some key strategies include:

Storytelling: Sharing personal stories and experiences can create powerful emotional connections and foster empathy among audiences.
Visual media: Images and videos can vividly portray the human impact of an issue, evoking empathy and inspiring action.
Community organizing: Building relationships and fostering dialogue within communities can create a sense of shared purpose and encourage collective action.

Empathy in Building Coalitions and Fostering Dialogue

Empathy is crucial for building bridges between different groups and fostering constructive dialogue. Activists can leverage empathy to:

Challenge stereotypes and prejudices: By humanizing those affected by an issue, activists can break down stereotypes and encourage understanding.
Promote active listening: Creating spaces for open communication and actively listening to diverse perspectives can foster mutual respect and collaboration.
Find common ground: Identifying shared values and goals can unite disparate groups and build powerful coalitions for change.

Examples of Activists Utilizing Empathy

History is replete with examples of activists who have effectively used empathy to inspire change and achieve remarkable results. Some notable figures include:

Nelson Mandela: Mandela's unwavering commitment to justice and reconciliation, even towards his former oppressors, exemplifies the power of empathy to overcome division and build a more inclusive society.
Malala Yousafzai: Malala's courage and empathy for girls denied access to education have galvanized global support for girls' education and empowered countless individuals to advocate for change.
Greta Thunberg: Thunberg's passionate advocacy for climate action, fueled by her empathy for future generations, has mobilized millions of young people worldwide to demand urgent action on climate change.
Empathy in the Digital Age

In the digital age, social media and online platforms have become powerful tools for activism. Activists can leverage these platforms to:

Share stories and information: Social media allows for the rapid dissemination of information and personal stories, creating opportunities to connect with and mobilize large audiences.
Build online communities: Online platforms can facilitate the formation of virtual communities, fostering dialogue, and encouraging collective action.
Organize virtual protests and campaigns: Digital tools enable activists to organize virtual events and campaigns, transcending geographical boundaries and amplifying their message.

Challenges and Considerations

While empathy is a powerful tool for activism, it is essential to acknowledge potential challenges and ethical considerations:

Empathy fatigue: Constant exposure to suffering can lead to emotional exhaustion and burnout. Activists need to prioritize self-care and establish healthy boundaries.
Empathy bias: Our empathy may be more readily extended to those who are similar to us or those whose stories resonate with our own experiences. Activists need to be mindful of these biases and actively work to cultivate empathy for all those affected by injustice.
Manipulation and exploitation: Empathy can be exploited for manipulative purposes. Activists need to ensure that their use of empathy is ethical and authentic, prioritizing the well-being and agency of those they seek to help.
Conclusion

Empathy is a driving force in activism, inspiring individuals to take action, mobilizing support, and fostering dialogue and collaboration. Activists can leverage empathy to connect with others, challenge stereotypes, and build powerful coalitions for change. In the digital age, online platforms offer new opportunities for activists to harness the power of empathy and create meaningful social change on a global scale. However, it is crucial to remain mindful of the potential challenges and ethical considerations associated with empathy in activism, ensuring that its use is authentic, responsible, and ultimately contributes to a more just and equitable world.

6.4 Empathy and Restorative Justice: Healing and Reconciliation

Traditional justice systems often focus on punishment and retribution, perpetuating cycles of harm and failing to address the root causes of conflict. Restorative justice, an approach grounded in empathy, offers an alternative path towards healing, accountability, and reconciliation in the aftermath of harm or conflict. This section explores how empathy-based approaches to justice can promote healing and build a more just and peaceful society.

Understanding Restorative Justice

Restorative justice shifts the focus from punishment to repairing harm and restoring relationships. It emphasizes:

Victim needs and healing: Prioritizes addressing the needs of those harmed, providing them with support, validation, and a voice in the process.
Offender accountability: Encourages offenders to take responsibility for their actions, understand the impact of their harm, and make amends.
Community involvement: Engages the community in supporting both victims and offenders and fostering a sense of collective responsibility for healing and reconciliation.
The Role of Empathy in Restorative Justice

Empathy is fundamental to the restorative justice process. It enables:

Understanding the impact of harm: Allows victims, offenders, and community members to understand the emotional, social, and physical consequences of the harm caused.
Fostering accountability: Helps offenders develop empathy for the victim and recognize the impact of their actions, promoting genuine remorse and a willingness to make amends.
Facilitating dialogue and understanding: Creates a safe space for open communication, allowing victims and

offenders to share their perspectives and work towards mutual understanding.
Promoting forgiveness and reconciliation: Encourages empathy and compassion, facilitating forgiveness, healing, and the restoration of relationships.
Examples of Restorative Justice Practices

Restorative justice practices are being implemented in various settings, including:

Schools: Peer mediation programs, restorative circles, and victim-offender conferences help address conflicts and bullying, fostering a more positive school climate.
Communities: Community conferencing brings together victims, offenders, and community members to discuss the harm caused and collectively determine appropriate ways to repair it.
Criminal justice system: Victim-offender dialogues, restorative justice courts, and community-based programs offer alternatives to traditional incarceration, focusing on rehabilitation and reintegration.
Benefits of Restorative Justice

Research suggests that restorative justice can lead to:

Reduced recidivism: Offenders who participate in restorative justice programs are less likely to re-offend compared to those who go through traditional court processes.
Increased victim satisfaction: Victims report higher levels of satisfaction with restorative justice processes, feeling more empowered and heard.
Improved community safety: Restorative justice strengthens community bonds and addresses the root causes of conflict, contributing to safer and more peaceful communities.

Enhanced emotional well-being: Restorative justice promotes healing and emotional recovery for both victims and offenders, reducing feelings of anger, fear, and resentment.

Challenges and Considerations

Implementing restorative justice effectively requires careful consideration of various factors:

Victim safety and empowerment: Ensuring victim safety and agency is paramount. Restorative justice should never pressure victims to participate or forgive.

Offender readiness: Offenders must be willing to take responsibility for their actions and genuinely engage in the process.

Community capacity: Restorative justice requires trained facilitators and community support to be successful.

Suitability for all cases: Restorative justice may not be appropriate for all types of offenses, particularly those involving serious violence or power imbalances.

Emerging Trends in Restorative Justice

Technology-enabled restorative justice: Online platforms and virtual conferencing tools are being used to facilitate restorative justice processes, increasing accessibility and overcoming geographical barriers.

Trauma-informed restorative justice: Recognizing the impact of trauma on both victims and offenders, trauma-informed approaches prioritize safety, empowerment, and sensitivity.

Restorative justice in addressing systemic issues: Restorative justice principles are being applied to address systemic issues like racism, discrimination, and historical injustices, promoting healing and reconciliation at a societal level.

Conclusion

Empathy-based approaches to justice, like restorative justice, offer a powerful alternative to traditional punitive systems. By prioritizing healing, accountability, and reconciliation, restorative justice can help break cycles of harm and build a more just and peaceful society. It fosters empathy, facilitates dialogue, and empowers communities to address conflict constructively. As restorative justice continues to evolve and adapt, it holds the potential to transform our approach to justice and contribute to a more compassionate and equitable world.

Empathy in the Digital Age

7.1 Empathy and Online Communication: Navigating the Nuances of Virtual Interaction

Introduction

Empathy, the ability to understand and share the feelings of others, is a cornerstone of meaningful human connection. In the digital age, where communication increasingly occurs online, the nuances of empathetic interaction are both challenged and transformed. This section delves into the complexities of empathy in online communication, exploring the hurdles it faces, the unique opportunities it presents, and the strategies we can employ to foster genuine connection in the virtual realm.

Challenges to Empathy in Online Communication

Absence of Nonverbal Cues: Face-to-face conversations are rich with nonverbal cues like facial expressions, body language, and tone of voice, which provide crucial context for understanding emotions. Online, these cues are often absent or significantly reduced, making it difficult to accurately gauge another person's emotional state. This can lead to misinterpretations, misunderstandings, and a diminished sense of connection.
Text-Based Misinterpretations: Written communication, especially in text messages or short online posts, is prone to misinterpretation. Sarcasm, humor, and complex emotions can be easily lost in translation, potentially leading to unintended offense or hurt feelings.
Physical Distance and Anonymity: The physical distance inherent in online communication can sometimes create a sense of detachment and reduce feelings of accountability. Anonymity, often afforded by online

platforms, can further disinhibit behavior, leading to increased negativity and reduced empathy.
Information Overload and Attention Deficit: The constant influx of information and stimuli in the digital world can lead to attention deficit and a decreased capacity for deep engagement. This can make it challenging to truly focus on and empathize with others online.
Opportunities for Empathy in Online Communication

Connecting Across Geographical Barriers: Online communication transcends geographical limitations, allowing us to connect with people from diverse backgrounds and cultures. This exposure to different perspectives can foster empathy and broaden our understanding of the world.
Increased Accessibility: Online platforms can provide a safe space for individuals who may struggle with social anxiety or face barriers to in-person communication. These platforms can empower them to express themselves and connect with others in a way that feels comfortable.
Amplifying Marginalized Voices: Social media and online forums can give a voice to marginalized communities and individuals who may not have access to traditional media outlets. This can raise awareness of social issues and foster empathy for those who are often overlooked.
Cultivating Asynchronous Connection: Online communication allows for asynchronous interaction, giving individuals time to reflect and thoughtfully respond to others. This can lead to more meaningful and empathetic exchanges.
Fostering Empathy in Virtual Interactions

Mindful Language: Choosing words carefully and considering their potential impact is essential for empathetic online communication. Avoiding inflammatory language, using inclusive terminology, and expressing gratitude can all contribute to a more positive and understanding online environment.

Active Listening: Even in the absence of nonverbal cues, active listening is crucial for conveying empathy online. Asking clarifying questions, summarizing to ensure understanding, and responding thoughtfully demonstrate genuine interest and care.

Emotional Awareness: Cultivating self-awareness and recognizing one's own emotions is key to empathizing with others online. Taking time to reflect on how our words might be perceived and being mindful of our emctional state can help us communicate more effectively and empathetically.

Utilizing Visual Cues: Emojis, GIFs, and other visual cues can help to convey emotion and add context to online communication. However, it's important to use them thoughtfully and ensure they align with the intended message.

Creating Safe Spaces: Fostering online communities with clear guidelines and moderation can help to create safe spaces where individuals feel comfortable expressing vulnerability and empathy.

Examples and Case Studies

Empathetic Email Communication: When responding to a customer complaint via email, taking the time to acknowledge their frustration, apologize for any inconvenience, and offer a solution demonstrates empathy and can help to de-escalate the situation.

Using Emojis Effectively: In a text message expressing condolences, including a simple heart emoji can convey warmth and compassion that might be lost in a purely text-based message.

Building Inclusive Online Communities: Online forums dedicated to specific health conditions or social identities can provide a supportive space for individuals to connect with others who understand their experiences, fostering empathy and reducing feelings of isolation.

Conclusion

Empathy in online communication requires conscious effort and a willingness to navigate the unique challenges of the digital realm. By embracing mindful language, active listening, and emotional awareness, we can foster genuine connection and understanding in our virtual interactions. As we continue to evolve in the digital age, cultivating empathy online will be essential for building a more compassionate and inclusive world.

7.2 Empathy and Social Media Use: The Double-Edged Sword of Connection

Introduction

Social media has become an undeniable force in modern society, transforming the way we connect, communicate, and consume information. Its impact on empathy, however, is complex and multifaceted. While social media can be a powerful tool for fostering empathy by connecting people across distances and raising awareness of social issues, it also has the potential to hinder empathy through online negativity, cyberbullying, and the creation of echo chambers. This section explores the intricate relationship between empathy and social media, examining both its positive and negative influences, and offering strategies for using social media in a way that promotes empathy and avoids negativity.

How Social Media Can Promote Empathy

Connecting Across Distances: Social media platforms allow us to connect with people from all over the world,

fostering cross-cultural understanding and empathy for those with different backgrounds and experiences.
Raising Awareness of Social Issues: Social media has become a powerful platform for raising awareness of social issues, from human rights violations to environmental concerns. By sharing stories and information, social media can help to generate empathy and inspire action.
Facilitating Support Networks: Online communities and support groups on social media can provide a sense of belonging and connection for individuals facing challenges, fostering empathy and reducing feelings of isolation.
Sharing Personal Experiences: Social media allows individuals to share their personal experiences, vulnerabilities, and struggles, fostering empathy and understanding among those who can relate.
How Social Media Can Hinder Empathy

Online Negativity and Cyberbullying: The anonymity and disinhibition often associated with online interactions can lead to increased negativity, cyberbullying, and online harassment. These experiences can erode empathy and contribute to a hostile online environment.
Echo Chambers and Polarization: Social media algorithms often create echo chambers, where users are primarily exposed to information and perspectives that align with their own. This can lead to increased polarization, reduced exposure to diverse viewpoints, and a decline in empathy for those who hold different beliefs.
Comparison and Competition: Social media can foster a culture of comparison and competition, where individuals present idealized versions of themselves. This can lead to feelings of inadequacy, envy, and a diminished capacity for empathy.
Information Overload and Desensitization: The constant barrage of information and emotionally charged content on social media can lead to information overload and

desensitization, making it difficult to process and empathize with the suffering of others.

Designing Social Media Platforms to Encourage Empathy

Promoting Positive Interactions: Social media platforms can be designed to encourage positive interactions by rewarding prosocial behavior, highlighting positive content, and providing tools for users to report and address negativity.

Encouraging Diverse Perspectives: Algorithms can be adjusted to expose users to a wider range of perspectives and viewpoints, breaking down echo chambers and fostering empathy for those with different beliefs.

Facilitating Meaningful Connections: Social media platforms can facilitate meaningful connections by promoting online communities and groups based on shared interests, values, or experiences.

Promoting Media Literacy: Social media platforms can play a role in promoting media literacy by providing users with tools and resources to critically evaluate information and identify misinformation.

Strategies for Using Social Media Empathetically

Practice Mindful Consumption: Be mindful of the content you consume on social media and limit exposure to negativity and emotionally draining content.

Engage in Thoughtful Interactions: When interacting with others online, practice active listening, choose your words carefully, and avoid making assumptions or judgments.

Cultivate Diverse Connections: Connect with people from different backgrounds and perspectives to broaden your understanding of the world and foster empathy.

Use Social Media for Good: Use social media to raise awareness of social issues, support causes you care about, and connect with others who share your values.

Examples and Case Studies

#MeToo Movement: The #MeToo movement, which gained momentum on social media, empowered individuals to share their experiences of sexual harassment and assault, fostering empathy and solidarity among survivors.
Social Media and Disaster Relief: Social media has played a crucial role in coordinating disaster relief efforts, connecting those in need with those who can offer assistance, and fostering empathy for those affected by tragedy.
The Impact of Cyberbullying: Studies have shown that cyberbullying can have a significant negative impact on mental health, leading to increased anxiety, depression, and even suicidal ideation.
Conclusion

Social media presents both opportunities and challenges for empathy in the digital age. By understanding the complex relationship between social media and empathy, we can harness its power for good, using it to connect with others, raise awareness of social issues, and build a more compassionate world. By practicing mindful consumption, engaging in thoughtful interactions, and cultivating diverse connections, we can use social media in a way that promotes empathy and avoids negativity.

7.3 Empathy and Digital Technology Design: Creating Human-Centered Technology

Introduction

In our increasingly digital world, technology has become deeply intertwined with nearly every aspect of our lives. From the smartphones we use to communicate to the software that powers our workplaces, technology's influence is undeniable. Yet, as technology advances at an unprecedented pace, it's crucial to ensure that it serves human needs and enhances our well-being. This is where empathy in digital technology design comes into play. By incorporating empathy into the design process, we can create technology that is not only functional and efficient but also user-friendly, accessible, and inclusive.

The Role of Empathy in Human-Centered Design

Human-centered design (HCD) is an approach to problem-solving that puts people at the center of the design process. 1 It emphasizes understanding the needs, emotions, and perspectives of users to create solutions that are truly relevant and effective. Empathy is the cornerstone of HCD, as it allows designers to step into the shoes of their users and understand their experiences on a deeper level.

Key Principles of Empathetic Technology Design

User Research: Conducting thorough user research is essential for understanding the needs, motivations, and pain points of users. This involves observing users in their natural environment, conducting interviews, and gathering feedback through surveys and usability testing.
Accessibility: Designing technology that is accessible to everyone, regardless of their physical or cognitive abilities, is a crucial aspect of empathetic design. This includes considering factors such as visual impairments, hearing loss, and motor limitations.
Inclusivity: Empathetic design strives to create technology that is inclusive of all users, regardless of their background, culture, or identity. This involves considering diverse perspectives and avoiding biases in the design process.
Emotional Design: Emotional design focuses on creating technology that evokes positive emotions and enhances

user well-being. This involves considering factors such as aesthetics, user experience, and the overall emotional impact of the technology.
Ethical Considerations: Empathetic design takes into account the ethical implications of technology, ensuring that it is used responsibly and does not contribute to harm or discrimination.
Examples of Empathetic Technology Design

Assistive Technologies: Assistive technologies, such as screen readers, voice recognition software, and prosthetic limbs, are designed to empower individuals with disabilities and enhance their quality of life.
Personalized Learning Technologies: Educational technologies that personalize learning experiences to individual student needs can help to improve engagement and academic outcomes.
Mental Health Apps: Mental health apps that provide support, resources, and coping mechanisms can help individuals manage their mental well-being and reduce stigma.
Inclusive Design in Gaming: Video games that offer customizable difficulty levels, diverse character representation, and accessible control options can create a more inclusive and enjoyable gaming experience for all.
User-Friendly Interfaces: Websites and applications with clear navigation, intuitive layouts, and accessible language can improve user experience and reduce frustration.
Case Studies

Prosthetics with Sensory Feedback: Advances in prosthetic technology have led to the development of prosthetic limbs that can provide sensory feedback, allowing users to feel sensations such as pressure and temperature. This technology has significantly improved the quality of life for amputees.

Duolingo's Language Learning App: Duolingo's language learning app uses gamification and personalized learning paths to make language learning engaging and accessible to a wide range of users.
Be My Eyes App: The Be My Eyes app connects visually impaired individuals with sighted volunteers who can assist them with everyday tasks through a live video call. This technology fosters empathy and empowers those with visual impairments.
The Future of Empathetic Technology Design

As technology continues to evolve, the role of empathy in design will become increasingly important. Emerging technologies such as artificial intelligence (AI) and virtual reality (VR) present both opportunities and challenges for empathetic design. AI-powered systems can be used to personalize user experiences and provide tailored support, while VR can create immersive experiences that foster empathy and understanding. However, it's crucial to ensure that these technologies are developed and used ethically, with a focus on human well-being.

Conclusion

Empathy is essential for creating technology that truly serves human needs and enhances our lives. By incorporating empathy into the design process, we can create technology that is user-friendly, accessible, inclusive, and ethically sound. As technology continues to advance, empathetic design will play a crucial role in shaping a future where technology empowers and connects us all.

7.4 Empathy and Artificial Intelligence: The Ethical Considerations of Emotionally Intelligent Machines

Introduction

Artificial intelligence (AI) is rapidly advancing, moving beyond simple task automation to more complex realms of human interaction. One of the most fascinating and potentially transformative developments is the emergence of emotionally intelligent AI – machines capable of understanding, interpreting, and even responding to human emotions. While this presents exciting possibilities, it also raises profound ethical questions about the role of empathy in AI and the potential consequences of creating machines that can understand and manipulate human feelings.

The Potential Benefits of Emotionally Intelligent AI

Enhanced Human-Computer Interaction: Emotionally intelligent AI can create more natural and intuitive interactions between humans and machines. This could lead to more personalized and user-friendly technology, from virtual assistants that understand our moods to educational software that adapts to our learning styles.
Improved Mental Health Care: AI-powered chatbots and virtual therapists could provide accessible and personalized mental health support, offering a safe space for individuals to express their emotions and receive guidance.
More Effective Customer Service: AI systems that can understand customer emotions can provide more empathetic and effective customer service, leading to increased satisfaction and loyalty.
Enhanced Education and Training: AI tutors that can recognize and respond to student emotions can personalize learning experiences and provide tailored

support, improving engagement and academic outcomes.

Social Good Applications: Emotionally intelligent AI could be used to detect and prevent cyberbullying, identify individuals at risk of self-harm, and provide support to those in crisis.

The Ethical Risks of Emotionally Intelligent AI

Manipulation and Exploitation: AI systems that can understand and respond to human emotions could be used to manipulate individuals for commercial or political gain. This raises concerns about the potential for AI to be used in targeted advertising, propaganda, and even political manipulation.

Privacy Violations: Emotionally intelligent AI often relies on collecting and analyzing personal data, including sensitive information about our emotions and mental states. This raises concerns about privacy violations and the potential for misuse of this data.

Bias and Discrimination: AI systems are trained on data, and if that data reflects existing biases, the AI system may perpetuate and even amplify those biases. This could lead to AI systems that discriminate against certain groups or individuals based on their emotional expressions.

Erosion of Human Connection: Over-reliance on emotionally intelligent AI could lead to a decline in genuine human connection and empathy. If we become accustomed to interacting with machines that simulate empathy, we may lose the ability to connect with and understand the emotions of other humans.

Job Displacement: As emotionally intelligent AI becomes more sophisticated, it could potentially replace human workers in fields that require emotional intelligence, such as customer service, therapy, and education.

Developing Ethical Guidelines for Emotionally Intelligent AI

Transparency and Explainability: AI systems should be transparent and explainable, so that users understand how they work and how their emotions are being interpreted.
Data Privacy and Security: Strict regulations should be in place to protect the privacy and security of user data collected by emotionally intelligent AI.
Fairness and Non-discrimination: AI systems should be designed and trained to be fair and non-discriminatory, ensuring that they do not perpetuate or amplify existing biases.
Human Oversight and Control: Human oversight and control should be maintained over emotionally intelligent AI systems, particularly in sensitive areas such as healthcare and education.
Continuous Monitoring and Evaluation: The impact of emotionally intelligent AI should be continuously monitored and evaluated to ensure that it is being used ethically and responsibly.
Examples and Case Studies

Affective Computing: Researchers are developing AI systems that can recognize and respond to human emotions through facial expressions, voice tone, and physiological signals. These systems have potential applications in healthcare, education, and human-computer interaction.
AI-Powered Chatbots for Mental Health: Woebot and other AI-powered chatbots are being used to provide mental health support and companionship, offering a safe space for individuals to express their emotions and receive guidance.
Ethical Concerns in AI-Powered Hiring: AI systems are being used in hiring processes to analyze candidate emotions and predict job performance. This raises concerns about bias, discrimination, and the potential for AI to perpetuate existing inequalities.

Conclusion

The development of emotionally intelligent AI presents both exciting possibilities and profound ethical challenges. By carefully considering the potential benefits and risks, and by developing ethical guidelines for its development and use, we can ensure that emotionally intelligent AI is used responsibly and for the betterment of humanity. It is crucial to prioritize human well-being, privacy, and fairness as we navigate this new frontier of technology.

Empathy and Environmentalism

8.1 Empathy and Nature Connectedness: Fostering a Sense of Belonging to the Natural World

Introduction

In an era marked by increasing urbanization and technological dependence, the concept of nature connectedness has gained significant attention. Nature connectedness refers to the subjective sense of belonging to the natural world, encompassing feelings of kinship, awe, and wonder towards nature. This connection is not merely about physical proximity to nature but also involves emotional, cognitive, and spiritual dimensions. Research has consistently demonstrated that individuals with a strong nature connection tend to exhibit greater pro-environmental attitudes and behaviors. Empathy, the ability to understand and share the feelings of others, plays a crucial role in fostering this connection.

The Interplay of Empathy and Nature Connectedness

Empathy and nature connectedness are intertwined in a reciprocal relationship. Empathy allows us to perceive and appreciate the intricate web of life, recognizing the interdependence of all living beings. When we connect with nature, we become more attuned to the needs of other species and the delicate balance of ecosystems. This understanding can evoke feelings of care and responsibility towards the natural world, motivating us to protect and preserve it.

Psychological Benefits of Nature Exposure

Spending time in nature has been shown to have numerous psychological benefits. Research indicates that nature exposure can reduce stress, improve mood, and enhance cognitive function. The sights, sounds, and smells of nature can have a calming effect on the mind, reducing anxiety and promoting relaxation. Studies have also shown that exposure to natural environments can improve attention, memory, and creativity.

Cultivating Awe and Wonder

Nature experiences can foster a sense of awe and wonder, leading to a deeper appreciation of the natural world. Awe is an emotion that arises when we encounter something vast and mysterious, something that transcends our ordinary understanding. When we witness the grandeur of a mountain range, the intricate patterns of a spider web, or the vastness of the ocean, we are filled with a sense of wonder and humility. These experiences can shift our perspective, reminding us that we are part of something larger than ourselves.

Promoting Pro-environmental Attitudes

Nature connectedness has been linked to a range of pro-environmental attitudes and behaviors. Individuals who feel connected to nature are more likely to engage in environmentally responsible actions, such as recycling, conserving energy, and supporting environmental causes. This connection can also inspire a sense of stewardship, motivating individuals to protect and preserve the natural world for future generations.

Examples and Case Studies

Forest Bathing: The practice of forest bathing, or shinrin-yoku, originated in Japan and involves immersing oneself in the sights, sounds, and smells of a forest. Studies have shown that forest bathing can reduce stress hormones, lower blood pressure, and boost the immune system.
Nature-based Therapy: Nature-based therapy programs utilize nature experiences to promote mental and emotional well-being. These programs have been shown to be effective in treating a range of conditions, including depression, anxiety, and PTSD.
Environmental Education: Environmental education programs that incorporate outdoor experiences can foster nature connectedness and promote pro-environmental attitudes in children and adults.
Conclusion

Empathy and nature connectedness are essential for cultivating a sense of belonging to the natural world and promoting pro-environmental attitudes. By spending time in nature, observing natural beauty, and appreciating the interconnectedness of all living things, we can enhance our empathy and deepen our connection to the natural world. This connection can inspire us to act as stewards of the planet, protecting and preserving it for future generations.

8.2 Empathy and Environmental Concern: Motivating Action to Protect the Planet

Introduction

Environmental concern, the apprehension and worry about environmental problems, is a critical driver of pro-environmental behavior. While scientific evidence and rational arguments play a role in raising awareness, empathy can be a powerful motivator for action. Empathy allows us to connect with the suffering of others, including non-human beings and future generations, fostering a sense of responsibility to protect the planet.

Empathy's Influence on Environmental Action

Empathy can motivate individuals to care about the environment in several ways:

Connecting with the Suffering of Others: Witnessing the impact of environmental degradation on human communities, such as those displaced by climate change or exposed to pollution, can evoke empathy and spur action. Similarly, learning about the suffering of animals due to habitat loss or pollution can trigger compassionate concern and a desire to help.
Recognizing Interdependence: Understanding the interconnectedness of all living things can foster a sense of responsibility for the well-being of the planet. Recognizing that our actions have consequences for others, both human and non-human, can motivate us to make more sustainable choices.
Envisioning Future Impacts: Empathy allows us to imagine the potential consequences of environmental degradation for future generations. This perspective can inspire a sense of intergenerational responsibility, motivating us to act now to protect the planet for those who will inherit it.
The Role of Empathy in Environmental Activism

Empathy plays a crucial role in environmental activism and advocacy. Activists often use stories, images, and personal narratives to evoke empathy and inspire action. By sharing the experiences of those affected by environmental problems, they can connect with the emotions of their audience and motivate them to get involved.

Examples of Empathy-Driven Environmental Action

Climate Change Activism: The movement to address climate change is fueled by empathy for those already experiencing its impacts, such as communities facing rising sea levels or extreme weather events. Activists often share stories of these communities to raise awareness and demand action.
Wildlife Conservation: Efforts to protect endangered species are often driven by empathy for animals facing extinction. Conservation organizations use images and stories of these animals to connect with the public and inspire support for their work.
Environmental Justice Movements: Environmental justice movements advocate for the equitable distribution of environmental benefits and burdens. These movements are often rooted in empathy for communities disproportionately affected by pollution and environmental degradation.
Harnessing Empathy for Environmental Advocacy

To effectively utilize empathy in environmental advocacy:

Tell compelling stories: Personal narratives and anecdotes can create emotional connections and make environmental issues more relatable.
Use powerful imagery: Images and videos can evoke strong emotions and convey the urgency of environmental problems.
Highlight shared values: Emphasize shared values such as compassion, justice, and responsibility to connect with a broader audience.

Empower individuals: Provide opportunities for people to take action and make a difference, fostering a sense of agency and hope.

Conclusion

Empathy is a powerful motivator for environmental concern and action. By connecting with the suffering of others, recognizing our interdependence, and envisioning future impacts, we can cultivate a deeper sense of responsibility for the planet. Harnessing the power of empathy in environmental advocacy can inspire individuals to take action and work towards a more sustainable future.

8.3 Empathy and Sustainable Living: Making Conscious Choices for a Healthier Planet

Introduction

Sustainable living involves making conscious choices that minimize our environmental impact and promote the well-being of the planet. While knowledge about environmental issues is essential, empathy plays a crucial role in translating that knowledge into action. By considering the impact of our choices on others, both human and non-human, we can make more sustainable decisions about consumption, waste, and resource use.

Empathy's Influence on Sustainable Choices

Empathy can inform our sustainable living practices in several ways:

Reducing Consumption: Recognizing the environmental and social costs associated with excessive consumption can lead to more mindful purchasing decisions. Considering the labor conditions involved in production,

the resources used, and the waste generated can help us prioritize quality over quantity and choose products that align with our values.
Minimizing Waste: Empathy for future generations can motivate us to reduce waste and conserve resources. Understanding that our actions today will affect the availability of resources for future generations can inspire us to adopt practices like composting, recycling, and reducing single-use plastics.
Choosing Eco-friendly Products: Considering the impact of our choices on the environment and other species can lead us to choose eco-friendly products. For example, opting for products made from sustainable materials, supporting companies with ethical practices, and choosing plant-based options can reduce our environmental footprint.
Practical Tips for Sustainable Living

Reduce, Reuse, Recycle: Prioritize reducing consumption, finding creative ways to reuse items, and recycling whenever possible.
Conserve Energy: Use energy-efficient appliances, reduce reliance on private vehicles, and opt for renewable energy sources.
Choose Sustainable Food: Reduce meat consumption, prioritize locally sourced and organic produce, and minimize food waste.
Support Sustainable Businesses: Choose businesses that prioritize sustainability and social responsibility.
Minimize Water Use: Conserve water through efficient fixtures, responsible landscaping, and mindful water consumption.
Empathy and Corporate Social Responsibility

Empathy can also motivate businesses to adopt sustainable practices and prioritize social responsibility. Companies that consider the impact of their operations on the environment, their employees, and the communities they serve are more likely to make ethical and sustainable choices. This can include initiatives such as reducing emissions, supporting fair labor practices, and investing in community development.

Examples of Empathy-Driven Sustainable Practices

Zero Waste Movement: The zero waste movement encourages individuals to minimize waste generation through practices like composting, reusable containers, and mindful consumption.
Minimalism: Minimalism promotes a lifestyle focused on essential possessions and experiences, reducing consumption and environmental impact.
Plant-Based Diets: Choosing plant-based diets can significantly reduce one's environmental footprint due to the lower resource requirements of plant-based agriculture.
Sustainable Fashion: Supporting sustainable fashion brands that prioritize ethical production, fair labor practices, and eco-friendly materials.
Conclusion

Empathy plays a crucial role in motivating sustainable living practices. By considering the impact of our choices on others, both human and non-human, we can make more conscious decisions about consumption, waste, and resource use. Adopting sustainable practices not only benefits the environment but also promotes a more equitable and just society.

8.4 Empathy and Future Generations: Leaving a Legacy of Care and Responsibility

Introduction

Our actions today will have profound and lasting consequences for future generations. Empathy allows us to step outside our immediate concerns and consider the needs of those who will inherit the planet long after we are gone. This perspective can inspire a sense of intergenerational responsibility, motivating us to work towards a more sustainable and equitable future for all.

The Concept of Intergenerational Justice

Intergenerational justice is a principle that emphasizes the moral obligation of current generations to consider the well-being of future generations. It suggests that we have a duty to leave the planet in a condition that allows future generations to flourish and meet their own needs. This includes ensuring the availability of resources, protecting the environment, and promoting social equity.

Ethical Implications of Our Actions

Our current consumption patterns, resource use, and environmental impact have significant ethical implications for future generations. If we continue on a path of unsustainable development, we risk depleting resources, degrading the environment, and creating social inequalities that will burden future generations. Empathy challenges us to consider the long-term consequences of our actions and make choices that prioritize the well-being of future generations.

Challenges of Balancing Present and Future Needs

Balancing the needs of the present with the needs of the future can be challenging. Economic pressures, political considerations, and short-term thinking can often prioritize immediate gains over long-term sustainability. However, empathy can help us overcome these challenges by reminding us of our interconnectedness and our responsibility to those who will come after us.

Initiatives Prioritizing Future Generations

Several initiatives prioritize the well-being of future generations:

Sustainable Development Goals: The United Nations Sustainable Development Goals (SDGs) provide a framework for addressing global challenges, including poverty,1 inequality, and climate change, with a focus on intergenerational equity.

Climate Action Initiatives: Efforts to mitigate climate change, such as transitioning to renewable energy and promoting sustainable transportation, are crucial for protecting the planet for future generations.
Environmental Conservation: Protecting biodiversity, conserving natural resources, and restoring ecosystems are essential for ensuring the long-term health of the planet.
Education and Awareness: Educating future generations about sustainability and empowering them to take action is crucial for creating a more sustainable future.

Cultivating Intergenerational Empathy

To foster intergenerational empathy:

Encourage long-term thinking: Promote policies and practices that prioritize long-term sustainability over short-term gains.
Foster intergenerational dialogue: Create opportunities for dialogue and collaboration between generations to share perspectives and build understanding.
Promote education and awareness: Educate future generations about environmental issues, social justice, and the importance of intergenerational responsibility.
Empower youth: Provide young people with the tools and resources they need to become agents of change and create a more sustainable future.
Examples of Intergenerational Initiatives
Youth Climate Movements: Young people around the world are leading climate action movements, demanding that leaders take action to protect their future.
Intergenerational Community Projects: Projects that bring together different generations to work on common goals, such as community gardens or environmental restoration efforts, can foster intergenerational understanding and collaboration.
Legacy Letters: Writing letters to future generations expressing hopes and concerns for the future can be a powerful way to connect with future generations and inspire action.
Conclusion
Empathy for future generations is essential for creating a sustainable and equitable future. By considering the long-term consequences of our actions and prioritizing the well-being of those who will inherit the planet, we can leave a legacy of care and responsibility. Cultivating intergenerational empathy can empower us to work together to build a better future for all.

Empathy and Animal Welfare

9.1 Empathy and Animal Cognition: Recognizing the Sentience of Other Species

Introduction

The concept of sentience, the ability to experience feelings and sensations, has long been a subject of debate, particularly when it comes to non-human animals. Historically, animals were often viewed as unfeeling beings, driven purely by instinct. However, a growing body of research in animal cognition is challenging these traditional views, revealing the complex inner lives of animals and their capacity for emotions, intelligence, and social bonds. This chapter delves into the latest research on animal sentience, exploring how recognizing the capacity for feelings in other species can foster empathy, respect, and a more inclusive understanding of life on Earth.

The Rise of Animal Cognition

Animal cognition, the study of animal minds and their mental processes, has emerged as a vibrant field of research in recent decades. Scientists are employing innovative methods to explore animal intelligence, emotions, and social behavior, leading to remarkable discoveries that are reshaping our understanding of the animal kingdom.

Intelligence and Problem-Solving Abilities

Research has shown that animals possess remarkable problem-solving abilities, often demonstrating intelligence that rivals or even surpasses that of young children. For instance, New Caledonian crows have been observed crafting tools from leaves and twigs to extract insects from crevices, showcasing their ability for complex problem-solving and tool use. Similarly, octopuses, known for their intelligence, have been documented escaping from enclosures, solving mazes, and even recognizing individual human caretakers.

Emotions and Social Complexity

Animals also exhibit a wide range of emotions, including joy, grief, anger, and fear. Elephants, known for their strong social bonds, have been observed mourning the loss of family members, displaying behaviors such as touching the bones of deceased individuals and revisiting their graves. Similarly, studies on primates have revealed their capacity for empathy, cooperation, and even reconciliation after conflicts.

Communication and Language

Animal communication is another area of active research, with scientists discovering intricate communication systems in various species. Dolphins, for example, use a complex system of whistles and clicks to communicate with each other, while honeybees perform intricate dances to convey information about the location of food sources. Research on parrots has even shown their ability to understand and use human language, demonstrating their cognitive flexibility and capacity for learning.

Challenging Anthropocentric Views

The findings from animal cognition research challenge anthropocentric views that place humans at the center of the universe, considering other species as inferior. By recognizing the intelligence, emotions, and social complexity of animals, we can move towards a more inclusive understanding of sentience, acknowledging the intrinsic value of all living beings.

Fostering Empathy and Respect

Recognizing the sentience of other species is crucial for fostering empathy and respect for animals. When we understand that animals can feel pain, joy, and fear, just as we do, it becomes difficult to justify treating them as mere commodities or objects. This recognition can lead to a shift in our attitudes and behaviors towards animals, encouraging us to treat them with compassion and kindness.

Examples and Case Studies

Koko the Gorilla: Koko, a gorilla who learned sign language, demonstrated remarkable cognitive abilities and emotional depth. She expressed a wide range of emotions, including love, sadness, and even humor, challenging traditional views of animal intelligence and emotions.
Alex the Parrot: Alex, an African grey parrot, was the subject of groundbreaking research on animal language and cognition. He learned to identify objects, colors, and shapes, and even understood concepts like "bigger" and "smaller," demonstrating the cognitive abilities of birds.
The Cambridge Declaration on Consciousness: In 2012, a group of prominent neuroscientists signed the Cambridge Declaration on Consciousness, acknowledging that non-human animals, including mammals, birds, and even invertebrates like octopuses, possess the neurological substrates that generate consciousness.

Conclusion

The latest research on animal cognition provides compelling evidence that animals are sentient beings, capable of experiencing a wide range of emotions and engaging in complex cognitive processes. Recognizing the sentience of other species is essential for fostering empathy, respect, and a more ethical relationship with the animal kingdom. By understanding the inner lives of animals, we can move towards a more compassionate and inclusive world, where all living beings are treated with dignity and kindness.

9.2 Empathy and the Ethics of Animal Use: Rethinking Our Relationship with Animals

Introduction

Our relationship with animals is complex and multifaceted. We cherish them as companions, admire their beauty and grace, and yet, we also exploit them for our own needs and desires. This chapter delves into the ethical implications of our use of animals for food, entertainment, and research, exploring how empathy can inform our choices and encourage us to seek more compassionate alternatives.

The Ethical Concerns of Animal Use

The use of animals for human purposes raises a host of ethical concerns, particularly when it involves practices that cause suffering or deprive animals of their basic needs and freedoms.

Factory Farming: The vast majority of animals raised for food are confined in factory farms, where they endure cramped, unsanitary conditions, and are subjected to painful procedures such as beak trimming and tail docking. These practices raise serious ethical questions about the treatment of animals and the sustainability of our food system.

Animal Testing: Millions of animals are used in research and testing each year, often subjected to painful procedures and deprived of social interaction and natural behaviors. While animal research has led to medical advancements, the ethical cost of using animals in this way is increasingly being questioned.

Animals in Entertainment: Animals are used in various forms of entertainment, including circuses, zoos, and rodeos. These practices often involve confining animals to unnatural environments, forcing them to perform unnatural behaviors, and depriving them of their basic needs.

Empathy and Ethical Decision-Making

Empathy plays a crucial role in ethical decision-making, particularly when it comes to our treatment of animals. By putting ourselves in the place of another being, we can begin to understand their experiences and perspectives, leading to more compassionate choices.

Seeking Compassionate Alternatives

Empathy can also inspire us to seek alternatives to practices that harm animals. In the realm of food, this means exploring plant-based diets and supporting sustainable agriculture practices that prioritize animal welfare. In research, it means investing in alternative methods, such as in vitro testing and computer modeling, that can reduce or eliminate the need for animal experimentation. And in entertainment, it means supporting organizations that prioritize animal welfare and promote natural behaviors.

The Benefits of Compassionate Choices

Making compassionate choices about our use of animals not only benefits animals but also has positive implications for human health, the environment, and society as a whole. Plant-based diets, for instance, have been linked to lower risks of heart disease, type 2 diabetes, and certain types of cancer. Reducing our reliance on animal agriculture can also help mitigate climate change and conserve precious resources.

Examples and Case Studies

The rise of plant-based alternatives: The growing popularity of plant-based meats, dairy alternatives, and vegan restaurants reflects a shift in consumer preferences towards more compassionate and sustainable food choices.
The development of animal-free research methods: Scientists are making significant progress in developing alternative research methods that can replace animal testing, such as organ-on-a-chip technology and human-based clinical trials.
The transition to cruelty-free products: Many companies are now offering cruelty-free cosmetics and household products, reflecting a growing awareness of the ethical concerns surrounding animal testing.
Conclusion

Empathy can play a powerful role in rethinking our relationship with animals, encouraging us to make more compassionate choices about our use of animals for food, entertainment, and research. By seeking alternatives to practices that harm animals, we can create a more just and compassionate world for all living beings.

9.3 Empathy and Animal Advocacy: Speaking Up for Those Who Cannot Speak for Themselves

Introduction

Animals, unlike humans, cannot voice their suffering, express their needs, or advocate for their rights. They rely on us, as empathetic beings, to recognize their plight, understand their needs, and speak up on their behalf. This chapter explores the vital role of empathy in motivating individuals to become advocates for animal welfare and champions of animal rights. It examines how empathy can inspire action to protect animals from cruelty, exploitation, and indifference, and how it can fuel the movement for a more just and compassionate world for all living beings.

The Power of Empathy in Animal Advocacy

Empathy lies at the heart of animal advocacy. It is the driving force that compels individuals to take a stand against injustice, to challenge the status quo, and to demand a better world for animals. When we connect with the suffering of another being, when we feel their pain as our own, we are moved to act. This empathetic response is what fuels the animal advocacy movement.

Inspiring Action and Change

Empathy can inspire a wide range of actions in support of animal welfare. It can lead individuals to:

Raise awareness: Sharing information about animal welfare issues, educating others about the plight of

animals, and promoting compassionate choices can help create a more informed and empathetic society.
Support animal welfare organizations: Donating to or volunteering with organizations that rescue, rehabilitate, and advocate for animals can make a tangible difference in the lives of animals in need.
Lobby for stronger animal protection laws: Advocating for legislation that protects animals from cruelty, exploitation, and neglect can help create a legal framework that safeguards animal welfare.
Promote compassionate treatment of animals: Encouraging businesses, institutions, and individuals to adopt practices that prioritize animal welfare can help create a more compassionate society.
Make conscious consumer choices: Choosing cruelty-free products, supporting companies that prioritize animal welfare, and reducing or eliminating consumption of animal products can help drive demand for more ethical and sustainable practices.
The Role of Animal Welfare Organizations and Activists

Animal welfare organizations and activists play a crucial role in advocating for animals, often working tirelessly to expose cruelty, rescue animals in need, and promote animal rights. Their work is driven by empathy, a deep understanding of animal needs, and a commitment to creating a better world for animals.

Examples of Animal Advocacy in Action

The Humane Society of the United States (HSUS): HSUS is one of the largest animal welfare organizations in the world, working to protect animals through advocacy, education, and direct care.
People for the Ethical Treatment of Animals (PETA): PETA is a global animal rights organization known for its

provocative campaigns and investigations into animal cruelty.
Animal Equality: Animal Equality is an international organization that uses investigations, public education, and corporate outreach to expose and end cruelty to farmed animals.
The Jane Goodall Institute: Founded by renowned primatologist Jane Goodall, this institute works to protect chimpanzees and their habitats, promote conservation, and inspire action on behalf of all living beings.
The Impact of Empathy-Driven Advocacy

Empathy-driven advocacy has led to significant progress in animal welfare, including:

Increased awareness of animal welfare issues: Public awareness of issues such as factory farming, animal testing, and wildlife trafficking has increased significantly in recent years, thanks in part to the efforts of animal advocates.
Stronger animal protection laws: Many countries have strengthened their animal protection laws, banning practices such as dogfighting and bear bile farming, and increasing penalties for animal cruelty.
The rise of cruelty-free products and practices: The demand for cruelty-free products has grown significantly, leading many companies to adopt more ethical and sustainable practices.
Increased support for animal welfare organizations: Donations to and volunteerism with animal welfare organizations have increased, providing vital resources to support their work.
Conclusion

Empathy is a powerful force for change, motivating individuals to speak up for those who cannot speak for themselves. By harnessing the power of empathy, we can create a world where animals are treated with compassion, respect, and dignity.

9.4 Empathy and Human-Animal Interactions: Building Bonds of Mutual Respect

Introduction

Humans and animals have shared a close relationship for millennia. From the earliest days of domestication to the modern-day bond between humans and their companion animals, our interactions with animals have shaped our societies, cultures, and even our own evolution. This chapter explores how empathy can inform our interactions with animals, leading to more compassionate and respectful relationships that benefit both humans and animals.

The Role of Empathy in Human-Animal Interactions

Empathy is fundamental to building positive and mutually beneficial relationships with animals. By understanding and respecting their needs, emotions, and perspectives, we can create interactions that are enriching and fulfilling for both parties.

Responsible Pet Ownership

Empathy plays a crucial role in responsible pet ownership. It involves understanding the specific needs of different species, providing appropriate care, and respecting their natural behaviors. This includes:

Providing for their physical needs: Ensuring access to nutritious food, clean water, comfortable shelter, and appropriate veterinary care.

Meeting their emotional needs: Providing opportunities for socialization, play, and mental stimulation.
Respecting their boundaries: Recognizing and respecting their individual personalities and preferences, avoiding actions that cause fear or distress.
Understanding their communication: Learning to interpret their body language and vocalizations to better understand their needs and emotions.

Animal-Assisted Therapy

Animal-assisted therapy (AAT) harnesses the power of the human-animal bond to promote healing and well-being. AAT involves interacting with animals in a therapeutic setting to achieve specific goals, such as reducing stress, improving social skills, and increasing physical activity. Empathy is essential for AAT practitioners to:

Select appropriate animals: Choosing animals that are well-suited for therapeutic work, considering their temperament, training, and health.
Monitor animal welfare: Ensuring the animals involved in therapy are comfortable, relaxed, and not showing signs of stress or overwork.
Facilitate positive interactions: Creating a safe and supportive environment for both humans and animals, fostering positive and mutually beneficial interactions.

Humane Education

Humane education plays a vital role in fostering empathy and respect for animals. It involves teaching children and adults about animal welfare, responsible pet ownership, and the interconnectedness of all living beings. Empathy is central to humane education, as it:

Promotes understanding: Helps individuals understand the needs, emotions, and perspectives of animals.

Encourages compassion: Fosters compassion and respect for all living beings.
Inspires action: Empowers individuals to make a difference in the lives of animals.
The Benefits of Human-Animal Interactions

Positive human-animal interactions can have numerous benefits for both humans and animals. For humans, interacting with animals can:

Reduce stress and anxiety: Studies have shown that interacting with animals can lower levels of cortisol, a stress hormone.
Improve physical health: Pet ownership has been linked to lower blood pressure, cholesterol levels, and risk of heart disease.
Enhance social skills: Interacting with animals can help improve social skills, particularly in children and individuals with autism.
Increase emotional well-being: The companionship of animals can provide comfort, reduce loneliness, and promote feelings of happiness and well-being.
For animals, positive interactions with humans can:

Provide enrichment: Interactions with humans can provide mental and physical stimulation, enriching their lives.
Improve welfare: Positive interactions can contribute to improved animal welfare, particularly for animals in shelters or captive environments.
Strengthen the human-animal bond: Positive interactions can strengthen the bond between humans and animals, leading to more fulfilling relationships.
Examples and Case Studies

The Sanctuary at Best Friends Animal Society: This sanctuary in Kanab, Utah, provides a safe haven for thousands of rescued animals, demonstrating the power of compassion and the human-animal bond.

The Grey Muzzle Organization: This organization supports senior dogs in need, providing funding and resources to animal shelters and rescue groups.

Heifer International: This non-profit organization works to end hunger and poverty by providing livestock and training to communities in need, demonstrating the positive impact of human-animal partnerships.

Conclusion

Empathy is the cornerstone of positive and mutually beneficial human-animal interactions. By understanding and respecting the needs, emotions, and perspectives of animals, we can build relationships that enrich our lives and promote the well-being of all living beings.

Empathy in Conflict Resolution

10.1 Empathy and Conflict Escalation: Recognizing the Role of Misunderstanding and Bias

Empathy, the ability to understand and share the feelings of others, plays a crucial role in conflict resolution. Conversely, a lack of empathy can significantly contribute to conflict escalation. By fostering miscommunication, misinterpretations, and the perpetuation of stereotypes and biases, an absence of empathy can transform minor disagreements into major conflicts.

How Lack of Empathy Fuels Conflict Escalation

Miscommunication: When individuals fail to empathize with others, they may not accurately interpret their messages. This can lead to misunderstandings, misinterpretations, and a breakdown in communication, escalating the conflict.

Misinterpretations: Without empathy, individuals may attribute malicious intent to the actions of others, even when such intent is absent. These misinterpretations can fuel resentment and escalate the conflict.

Perpetuation of Stereotypes and Biases: A lack of empathy can reinforce stereotypes and biases. By failing to understand the perspectives and experiences of others, individuals may rely on preconceived notions, leading to discriminatory behavior and escalating the conflict.

How Empathy Can De-escalate Tense Situations

Promoting Understanding: Empathy allows individuals to see the situation from the other person's perspective. This understanding can help to de-escalate tense situations by reducing misunderstandings and misinterpretations.

Perspective-Taking: By encouraging perspective-taking, empathy helps individuals to appreciate the other person's feelings and motivations. This can lead to a more compassionate and understanding approach to conflict resolution.

Building Trust: Empathy can help to build trust between individuals. When people feel understood and valued, they are more likely to trust the other person and work towards a mutually acceptable solution.

Real-World Examples of Conflict Escalation Due to Lack of Empathy

Interpersonal Disputes: In personal relationships, a lack of empathy can lead to misunderstandings, hurt feelings, and resentment. For example, a partner who fails to empathize with their partner's stress may inadvertently escalate an argument.

Workplace Conflicts: In the workplace, a lack of empathy can contribute to a hostile work environment. For example, a manager who fails to empathize with their employees' workload may create a stressful and unproductive environment.

International Relations: On a global scale, a lack of empathy can fuel international conflicts. For example, a nation that fails to understand the historical grievances of another nation may engage in actions that escalate tensions.

How to De-escalate Conflict and Prevent Further Escalation

Recognize and Address Biases: Individuals should be mindful of their own biases and actively work to challenge them. This can help to prevent discriminatory behavior and promote understanding.

Actively Listen to Opposing Viewpoints: Active listening involves paying attention to the other person's perspective, asking clarifying questions, and reflecting on their message. This can help to reduce misunderstandings and build trust.

Seek Common Ground: Even in the midst of conflict, there is often common ground to be found. By focusing on shared interests and goals, individuals can work towards a mutually acceptable solution.

Conclusion

Empathy plays a critical role in conflict resolution. By promoting understanding, perspective-taking, and trust, empathy can help to de-escalate tense situations and prevent further escalation. Conversely, a lack of empathy can fuel conflict by contributing to miscommunication, misinterpretations, and the perpetuation of stereotypes and biases. By recognizing the importance of empathy and actively working to cultivate it, individuals can improve their conflict resolution skills and build stronger relationships.

10.2 Empathy and Mediation: Facilitating Communication and Understanding

Mediation is a structured process where a neutral third party, the mediator, assists disputing parties in reaching a mutually acceptable resolution. Empathy plays a pivotal role in effective mediation, facilitating communication, fostering understanding, and guiding the parties towards a resolution.

The Role of Empathy in Mediation

Facilitating Communication: Mediators use empathy to create a safe and comfortable environment where parties can openly express their perspectives and concerns. By demonstrating understanding and acknowledging the emotions of each party, mediators can facilitate effective communication and prevent misunderstandings.

Fostering Understanding: Empathy enables mediators to help parties see the situation from each other's perspectives. By encouraging perspective-taking and highlighting common ground, mediators can foster a deeper understanding of the underlying issues and motivations driving the conflict.

Guiding Towards Resolution: Empathetic mediators can guide parties towards a resolution by helping them identify their needs and interests, explore potential solutions, and negotiate mutually acceptable outcomes. By fostering a sense of collaboration and trust, mediators can empower parties to reach a resolution that addresses their underlying needs.

Skills and Qualities of an Effective Mediator

Empathy: The ability to understand and share the feelings of others is essential for effective mediation. Empathetic mediators can build rapport with the parties, facilitate communication, and foster understanding.

Active Listening: Active listening involves paying attention to both the verbal and nonverbal communication of the parties, asking clarifying questions, and reflecting on their messages. Active listening helps mediators to understand the perspectives of each party and identify their underlying needs and interests.

Impartiality: Mediators must remain impartial and neutral throughout the mediation process. This means avoiding taking sides or imposing their own opinions on the parties. Impartiality helps to build trust and ensures that the mediation process is fair and equitable.

Examples of Successful Mediation Cases

Family Disputes: Mediation can be effective in resolving family disputes, such as divorce, child custody, and inheritance disputes. By facilitating communication and fostering understanding, mediators can help families reach agreements that address their needs and protect the best interests of children.

Workplace Conflicts: Mediation can be used to resolve workplace conflicts, such as harassment, discrimination, and personality clashes. By creating a safe space for dialogue and encouraging perspective-taking, mediators can help employees resolve their differences anc restore a productive work environment.

Community Disputes: Mediation can be used to address community disputes, such as land use conflicts, environmental disputes, and neighborhood disputes. By bringing together diverse stakeholders and facilitating collaborative problem-solving, mediators can help communities reach mutually beneficial solutions.

The Different Stages of Mediation

Opening: The mediator sets the stage for the mediation process, explaining the ground rules and establishing a safe and comfortable environment.

Storytelling: Each party has the opportunity to share their perspective on the conflict, uninterrupted. The mediator uses active listening and empathy to understand the underlying needs and interests of each party.

Problem-Solving: The mediator facilitates a joint discussion to identify the core issues and explore potential solutions. Empathy helps the mediator to guide the parties towards a mutually acceptable outcome.

Agreement: If an agreement is reached, the mediator helps the parties to formalize the agreement in writing. The agreement should be clear, concise, and address the needs of all parties.

Conclusion

Empathy is a cornerstone of effective mediation. By facilitating communication, fostering understanding, and guiding the parties towards a resolution, empathetic mediators can help disputing parties reach mutually acceptable outcomes. By cultivating empathy, active listening skills, and impartiality, mediators can play a vital role in resolving conflicts and building stronger relationships.

10.3 Empathy and Negotiation: Finding Common Ground and Building Trust

Negotiation is a process where two or more parties with differing interests attempt to reach a mutually acceptable agreement. Empathy can be a powerful tool in negotiation, allowing parties to understand each other's needs, interests, and underlying motivations. By fostering understanding and building trust, empathy can facilitate compromise and lead to mutually beneficial agreements.

How Empathy Enhances Negotiation

Understanding Needs and Interests: Empathy allows negotiators to see beyond stated positions and understand the underlying needs and interests of the other party. This understanding can help to identify potential areas of agreement and create opportunities for mutually beneficial solutions.

Building Trust: Empathy fosters trust between negotiators. When parties feel understood and respected, they are more likely to be open to compromise and collaboration. Trust is essential for building strong relationships and reaching sustainable agreements.

Facilitating Compromise: Empathy can help negotiators to identify areas where they can compromise without sacrificing their core interests. By understanding the other party's priorities, negotiators can find creative solutions that address the needs of all parties involved.

Negotiation Strategies that Prioritize Empathy

Active Listening: Active listening involves paying attention to the other party's verbal and nonverbal communication, asking clarifying questions, and reflecting on their message. Active listening demonstrates respect and helps to build rapport, creating a more collaborative atmosphere.

Reframing: Reframing involves restating the other party's concerns in a way that highlights common ground and promotes understanding. Reframing can help to diffuse tension and shift the focus towards finding solutions.

Finding Common Ground: Identifying shared interests and goals can help to build a foundation for agreement. By focusing on common ground, negotiators can create a sense of collaboration and increase the likelihood of reaching a mutually beneficial outcome.

Examples of Empathy in Negotiation

Business Negotiations: In business negotiations, empathy can help to build relationships and create win-win solutions. For example, a supplier who understands the needs of their customer can offer flexible terms that benefit both parties.

Labor Negotiations: In labor negotiations, empathy can help to bridge the gap between management and labor. By understanding the concerns of both sides, negotiators can reach agreements that address the needs of employees while ensuring the company's continued success.

International Diplomacy: In international diplomacy, empathy can help to foster understanding and cooperation between nations. By recognizing the perspectives and interests of other countries, diplomats can work towards peaceful resolutions to conflicts.

Overcoming Obstacles and Building Rapport

Empathy can help negotiators to overcome obstacles and build rapport, even in challenging situations. By demonstrating understanding and respect, negotiators can create a more collaborative atmosphere and increase the likelihood of reaching a successful outcome.

Conclusion

Empathy is a valuable tool in negotiation. By fostering understanding, building trust, and facilitating compromise, empathy can lead to mutually beneficial agreements. By employing negotiation strategies that prioritize empathy, such as active listening, reframing, and finding common ground, negotiators can create a more collaborative atmosphere and increase the likelihood of reaching a successful outcome.

10.4 Empathy and Reconciliation: Healing the Wounds of Conflict

Conflict, whether interpersonal or international, often leaves deep wounds. These wounds can manifest as anger, resentment, mistrust, and even hatred. Reconciliation, the process of healing these wounds and restoring relationships, requires empathy. By fostering understanding, forgiveness, and compassion, empathy can help individuals and communities move beyond the pain of conflict and work towards a shared future.

The Role of Empathy in Reconciliation

Fostering Forgiveness: Empathy allows individuals to understand the perspectives and motivations of those who have harmed them. This understanding can facilitate forgiveness, a crucial step in the healing process. Forgiveness does not excuse the harm done, but it allows individuals to let go of anger and resentment, freeing them from the burden of the past.

Promoting Healing: Empathy helps individuals to acknowledge the pain and suffering of others, even those they may have considered enemies. This recognition of shared humanity can promote healing and create a sense of common ground. By understanding the impact of their actions on others, individuals can take responsibility for their role in the conflict and work towards making amends.

Building a Shared Future: Empathy can help communities to move beyond the divisions of the past and work towards a shared future. By fostering understanding and compassion, empathy can create a sense of shared identity and purpose. This can lead to collaborative efforts to rebuild communities and create a more just and peaceful society.

Examples of Empathy in Reconciliation

Truth and Reconciliation Commissions: Truth and reconciliation commissions (TRCs) are bodies tasked with investigating past human rights abuses and promoting healing and reconciliation. TRCs often provide a platform for victims to share their stories and for perpetrators to acknowledge their wrongdoing. By fostering empathy and understanding, TRCs can help communities to come to terms with the past and move forward.

Restorative Justice Practices: Restorative justice is an approach to justice that focuses on repairing the harm caused by crime and conflict. Restorative justice practices often involve bringing together victims, offenders, and community members to discuss the impact of the crime and develop a plan for making amends. 1 By fostering empathy and dialogue, restorative justice can promote healing and reconciliation.

Community Dialogue Initiatives: Community dialogue initiatives bring together people from different backgrounds to discuss sensitive issues and build relationships. These initiatives often use facilitated dialogue and storytelling to foster empathy and understanding. By creating a safe space for dialogue, community dialogue initiatives can help to break down stereotypes and build trust between different groups.

Challenges and Opportunities in Fostering Empathy in Post-Conflict Societies

Fostering empathy in post-conflict societies can be challenging. Trauma, mistrust, and deeply ingrained prejudices can make it difficult for people to understand and empathize with those they have been in conflict with. However, there are also opportunities for fostering empathy in post-conflict societies. By creating spaces for dialogue, promoting education about the conflict, and supporting initiatives that foster cross-cultural understanding, communities can work to build a more empathetic and reconciled society.

Conclusion

Empathy is essential for reconciliation. By fostering forgiveness, promoting healing, and building a shared future, empathy can help individuals and communities to move beyond the pain of conflict and work towards a more just and peaceful world. While fostering empathy in post-conflict societies can be challenging, it is also an opportunity to create a more compassionate and understanding world.

Empathy and Forgiveness

11.1 The Neurobiology of Forgiveness: Understanding the Brain's Role in Letting Go

Forgiveness, a complex and multifaceted process, involves a deliberate decision to release feelings of resentment or vengeance toward someone who has caused harm. While often viewed through a philosophical or spiritual lens, recent research has delved into the neurobiological underpinnings of forgiveness, revealing its profound impact on the brain and overall well-being.

The Neural Circuitry of Forgiveness

Several key brain regions play a crucial role in the intricate process of forgiveness:

Prefrontal Cortex: This executive control center of the brain is responsible for higher-order cognitive functions such as decision-making, emotional regulation, and empathy. In the context of forgiveness, the prefrontal cortex helps to override impulsive reactions, such as anger or revenge, and facilitates a more thoughtful and deliberate response.
Amygdala: This almond-shaped structure deep within the brain is the seat of emotions, particularly fear and anger. When someone wrongs us, the amygdala activates, triggering a cascade of physiological and emotional responses. Forgiveness involves dampening the activity of the amygdala, reducing the intensity of negative emotions.
Hippocampus: This seahorse-shaped structure plays a vital role in memory formation and retrieval. In the context of forgiveness, the hippocampus helps to reframe the memory of the transgression, shifting the focus away from

the harm caused and towards a more compassionate understanding of the offender's perspective.

Rewiring the Brain Through Forgiveness

The act of forgiveness can lead to significant neuroplasticity, essentially rewiring the brain in ways that promote well-being. Studies have shown that forgiveness can:

Decrease activity in brain regions associated with stress and negative emotions: By reducing the activity of the amygdala and other stress-related regions, forgiveness can help to alleviate anxiety, depression, and hostility.
Increase activity in regions associated with empathy and compassion: Forgiveness can enhance activity in the prefrontal cortex and other regions involved in empathy, promoting feelings of understanding and compassion towards the offender.
Strengthen neural connections associated with emotional regulation: Forgiveness can enhance the connectivity between the prefrontal cortex and the amygdala, improving the ability to regulate emotions and respond to challenging situations with greater composure.

Research on the Neural Correlates of Forgiveness

A growing body of research supports the notion that forgiveness has a profound impact on the brain. For example, a study published in the journal Psychological Science found that individuals who practiced forgiveness showed decreased activity in the amygdala and increased activity in the prefrontal cortex when recalling hurtful memories. Another study, published in NeuroImage, found that forgiveness was associated with increased connectivity between the prefrontal cortex and the amygdala, suggesting that forgiveness can strengthen the brain's ability to regulate emotions.

Examples of Forgiveness and Brain Changes

A study conducted at Stanford University found that participants who underwent an eight-week forgiveness training program showed significant reductions in stress hormones and improvements in mood compared to a control group.
Research on individuals who have suffered traumatic experiences, such as survivors of the Rwandan genocide, has shown that forgiveness can lead to decreased symptoms of post-traumatic stress disorder (PTSD) and improved psychological well-being.

Expert Insights

Dr. Frederic Luskin, Director of the Stanford Forgiveness Project, emphasizes the transformative power of forgiveness, stating, "Forgiveness is a process that allows you to let go of the negative emotions that are keeping you stuck in the past. It's not about condoning the behavior of the person who hurt you, but it's about freeing yourself from the pain and anger that are holding you back."

Conclusion

Forgiveness is not merely a moral or religious virtue; it is a neurobiological process that can profoundly impact the brain and overall well-being. By understanding the neural mechanisms involved in forgiveness, we can harness its power to heal emotional wounds, reduce stress, and promote greater peace and happiness in our lives.

11.2 Empathy as a Pathway to Forgiveness: Understanding the Offender's Perspective

Empathy, the ability to understand and share the feelings of others, serves as a powerful catalyst for forgiveness. By stepping into the shoes of those who have harmed us, we can gain a deeper understanding of their perspectives, motivations, and circumstances, fostering compassion and facilitating the process of letting go.

How Empathy Facilitates Forgiveness

Reduces Anger and Resentment: When we empathize with someone who has wronged us, we begin to see them as a complex individual with their own struggles and vulnerabilities. This perspective can help to diffuse anger and resentment, replacing them with a more nuanced understanding of their actions.

Diminishes the Desire for Revenge: Empathy can counteract the primal urge for revenge by highlighting the shared humanity between ourselves and the offender. Recognizing their capacity for suffering and remorse can diminish the desire to inflict pain in return.

Promotes Perspective-Taking: Empathy allows us to see the situation from the offender's point of view, considering factors that may have contributed to their harmful behavior. This broader perspective can help us to understand their actions, even if we do not condone them.

Fosters Compassion: By recognizing the offender's humanity and understanding their struggles, empathy can cultivate compassion, a key ingredient in the forgiveness process. Compassion allows us to see the offender as someone who deserves our understanding and kindness, even if they have made mistakes.

Examples of Empathy Leading to Forgiveness

Eva Kor: A Holocaust survivor who forgave the Nazi doctor who experimented on her and her twin sister. Kor's ability to empathize with Mengele's own troubled past allowed her to forgive him and find peace.
Azim Khamisa: After his son was tragically murdered, Khamisa reached out to the grandfather of the teenage killer, forging an unlikely friendship and working together to prevent youth violence.
Chris Williams: Paralyzed in a car accident caused by a drunk driver, Williams forgave the driver and became an advocate for restorative justice, emphasizing the importance of empathy and understanding in the healing process.

The Role of Empathy in Seeing the Humanity in Others

Empathy enables us to see beyond the labels of "victim" and "offender," recognizing the shared humanity that connects us all. Even those who have committed terrible acts are still human beings with their own stories, struggles, and capacity for change. Empathy allows us to see their humanity, fostering compassion and opening the door to forgiveness.

Expert Insights

Dr. Brené Brown, a research professor who studies courage, vulnerability, and shame, emphasizes the importance of empathy in forgiveness, stating, "Empathy is not about fixing, it's about listening and understanding. It's about connecting with the emotion that someone is experiencing, not trying to silver lining it."

Conclusion

Empathy serves as a crucial bridge to forgiveness, allowing us to understand the perspectives and motivations of those who have harmed us. By cultivating empathy, we can reduce anger, diminish the desire for revenge, and foster compassion, paving the way for healing and reconciliation.

11.3 Forgiveness and Interpersonal Relationships: Healing and Strengthening Bonds

Forgiveness plays a pivotal role in healing and strengthening interpersonal relationships, acting as a balm for wounds caused by conflict, betrayal, or misunderstandings. By fostering empathy and understanding, forgiveness can pave the way for reconciliation, restore trust, and deepen connections between individuals.

How Forgiveness Heals and Strengthens Relationships

Promotes Reconciliation: Forgiveness creates an opening for reconciliation, allowing individuals to move past the hurt and begin to rebuild their relationship. It signals a willingness to let go of the past and move forward together.
Restores Trust: Trust, a cornerstone of any healthy relationship, can be shattered by betrayal or dishonesty. Forgiveness, while not condoning the harmful behavior, can help to rebuild trust by demonstrating a commitment to the relationship and a belief in the other person's capacity for change.
Reduces Conflict: Lingering resentment and bitterness can fuel ongoing conflict and prevent resolution. Forgiveness can help to break this cycle, reducing the intensity of

negative emotions and creating space for constructive communication.
Fosters Empathy and Understanding: The act of forgiving often requires empathy, as individuals strive to understand the perspectives and motivations of those who have hurt them. This deeper understanding can foster compassion and strengthen the bond between them.
Enhances Emotional Intimacy: Forgiveness can create a sense of safety and vulnerability, allowing individuals to share their feelings and experiences more openly. This increased emotional intimacy can deepen the connection and strengthen the relationship.
Examples of Forgiveness in Relationships

Couples: In the aftermath of infidelity, forgiveness can be a long and challenging process, but it can ultimately lead to a stronger and more resilient relationship. Couples who successfully navigate forgiveness often report a renewed sense of commitment and a deeper appreciation for their partner.
Families: Forgiveness plays a crucial role in healing family rifts and restoring harmony. Whether it's forgiving a parent for past mistakes or reconciling with a sibling after a falling out, forgiveness can help to mend broken bonds and strengthen family ties.
Friendships: Forgiveness can help to repair friendships damaged by betrayal, misunderstandings, or hurtful words. By extending forgiveness, friends can demonstrate their commitment to the relationship and their willingness to move forward.
Empathy's Role in Creating Space for Forgiveness

Empathy is essential for creating a conducive environment for forgiveness in personal conflicts. By understanding the other person's perspective, we can begin to see their actions in a new light, recognizing their humanity and acknowledging their own struggles. This compassionate understanding can pave the way for forgiveness and reconciliation.

Expert Insights

Dr. John Gottman, a renowned relationship expert, emphasizes the importance of forgiveness in maintaining healthy relationships, stating, "Forgiveness is not about forgetting. It is about letting go of the resentment and bitterness that keep you stuck in the past."

Conclusion

Forgiveness serves as a powerful tool for healing and strengthening interpersonal relationships. By fostering empathy, restoring trust, and reducing conflict, forgiveness can pave the way for reconciliation and deepen connections between individuals. In the words of Desmond Tutu, "Forgiveness is not always easy. At times, it feels more painful than the wound we suffered, to forgive the one that inflicted it. And yet, there is no peace without forgiveness."

11.4 Forgiveness and Societal Healing: Reconciliation and Social Justice

Forgiveness, often viewed as a personal virtue, extends its transformative power to the societal level, playing a crucial role in promoting reconciliation and healing in communities and nations grappling with conflict, trauma, and historical injustices. By fostering empathy and understanding, forgiveness can pave the way for social justice, peacebuilding, and the creation of a more harmonious society.

The Role of Forgiveness in Societal Healing

Breaks the Cycle of Violence: In societies marked by conflict, cycles of violence and retribution often perpetuate division and suffering. Forgiveness can interrupt these cycles, offering an alternative to revenge and creating space for dialogue and reconciliation.
Promotes Truth and Reconciliation: Forgiveness is not about forgetting or condoning past wrongs but acknowledging them and seeking to understand their impact. Truth and reconciliation commissions, such as those in South Africa and Rwanda, provide a platform for victims to share their stories, perpetrators to acknowledge their actions, and communities to begin the process of healing.
Fosters Social Cohesion: Deep-seated divisions and mistrust can hinder societal progress and prevent communities from thriving. Forgiveness can help to bridge these divides, fostering empathy and understanding between different groups and promoting social cohesion.
Facilitates Peacebuilding: In post-conflict societies, forgiveness can be a critical component of peacebuilding efforts. By promoting reconciliation and fostering a sense of shared humanity, forgiveness can help to create a foundation for lasting peace.
Advances Social Justice: Forgiveness can be a powerful tool for addressing historical injustices and promoting social justice. By acknowledging past wrongs and seeking to make amends, societies can begin to heal and create a more equitable future.

Empathy's Role in Societal Forgiveness

Empathy is essential for facilitating forgiveness on a societal level. By understanding the perspectives and experiences of those who have been harmed, as well as the historical and social context that contributed to the conflict, we can cultivate compassion and create a more conducive environment for forgiveness and reconciliation.

Examples of Forgiveness and Societal Healing

Truth and Reconciliation Commissions: South Africa's Truth and Reconciliation Commission, established after the end of apartheid, provided a platform for victims to share their stories and perpetrators to confess their crimes in exchange for amnesty. This process, while imperfect, played a significant role in promoting healing and reconciliation in South Africa.
Restorative Justice Practices: Restorative justice focuses on repairing the harm caused by crime and conflict, bringing together victims, offenders, and community members to discuss the impact of the offense and determine how to make amends. This approach emphasizes accountability, healing, and the restoration of relationships.
Community Dialogue Initiatives: In communities divided by conflict or prejudice, dialogue initiatives can bring together people from different backgrounds to share their stories, build understanding, and foster empathy. These initiatives can help to break down stereotypes, promote reconciliation, and create a more inclusive society.
Expert Insights

Desmond Tutu, Archbishop Emeritus of Cape Town and a Nobel Peace Prize laureate, has been a vocal advocate for the power of forgiveness in societal healing, stating, "Without forgiveness, there is no future."

Conclusion

Forgiveness, while often viewed as a personal act, has profound implications for societal healing and reconciliation. By fostering empathy, promoting truth and reconciliation, and facilitating peacebuilding, forgiveness can help to create a more just and harmonious society. In the words of Nelson Mandela, "Forgiveness liberates the soul, it removes fear. That is why it is such a powerful weapon."

Empathy and Political Polarization

12.1 Empathy and Political Attitudes: Bridging the Divide

Introduction

In an era marked by increasing political polarization, the concept of empathy has emerged as a potential bridge across the chasm of ideological differences. Empathy, the ability to understand and share the feelings of others, can profoundly influence political attitudes and behaviors, shaping voting patterns, support for social policies, and the willingness to engage in constructive political dialogue. This section delves into the intricate relationship between empathy and political attitudes, exploring how empathy can mitigate prejudice, foster cooperation, and promote understanding across diverse ideologies.

Empathy's Impact on Political Attitudes and Behaviors

Recent research has illuminated the significant impact of empathy on various facets of political life:

Voting Patterns: Studies suggest that individuals with higher levels of empathy tend to lean towards political candidates who express compassion and concern for others. These individuals are more likely to vote for policies that address social inequalities and promote the well-being of marginalized groups.

Support for Social Policies: Empathy plays a crucial role in shaping attitudes towards social policies. Individuals with greater empathic capacity are more likely to support policies that benefit vulnerable populations, such as universal healthcare, affordable housing initiatives, and programs aimed at reducing poverty.

Willingness to Engage in Political Dialogue: Empathy fosters a greater willingness to engage in constructive political dialogue with those who hold differing views. By understanding and appreciating diverse perspectives, individuals can bridge political divides and work towards finding common ground.

Bridging Political Divides Through Empathy

Empathy has the potential to serve as a powerful antidote to political polarization. By cultivating empathy, indivicuals can:

Reduce Prejudice: Empathy can help to dismantle prejudice and stereotypes by encouraging individuals to see the world through the eyes of others. When people understand the experiences and challenges faced by those from different backgrounds, they are less likely to harbor negative biases.

Increase Support for Marginalized Groups: Empathy can lead to increased support for policies that benefit marginalized groups. By understanding the struggles of these communities, individuals are more likely to advocate for policies that promote equality and social justice.

Foster Constructive Dialogue: Empathy can facilitate constructive dialogue and cooperation across political differences. By approaching conversations with a willingness to understand opposing viewpoints, individuals can engage in respectful and productive discussions.

Examples of Empathy in Action

Research on Empathy and Political Ideology: A growing body of research demonstrates a correlation between empathy and political ideology. Studies have shown that individuals who score higher on measures of empathy tend to be more liberal in their political views. This may be due to the fact that empathy leads to a greater concern for social justice and equality.

Empathy-Driven Political Dialogue: Organizations such as the "National Institute for Civil Discourse" and the "Listen First Project" are dedicated to promoting empathy and constructive dialogue in political discourse. These initiatives bring together individuals from across the political spectrum to engage in respectful conversations and find common ground.

Conclusion

Empathy plays a pivotal role in shaping political attitudes and behaviors. By cultivating empathy, individuals can bridge political divides, reduce prejudice, and promote understanding across different ideologies. As political polarization continues to escalate, the importance of empathy in fostering a more cohesive and compassionate society cannot be overstated.

12.2 Empathy and Political Discourse: Promoting Respectful Communication

The Erosion of Civil Discourse

Political discourse, particularly in the digital age, has become increasingly fraught with hostility, incivility, and ad hominem attacks. The 24-hour news cycle, social media echo chambers, and the pursuit of sensationalism often prioritize conflict over constructive engagement. This toxic environment hinders productive dialogue and impedes progress on critical issues.

Empathy as a Counterforce

Empathy offers a powerful antidote to the negativity that pervades much of contemporary political discourse. By fostering understanding and respect, empathy can help to:

Reduce Hostility: When individuals make an effort to understand the perspectives and experiences of those with whom they disagree, they are less likely to resort to anger and aggression. Empathy encourages active listening and a willingness to consider alternative viewpoints.

Foster Understanding: Empathy enables individuals to see beyond stereotypes and recognize the shared humanity of those who hold different political beliefs. This recognition can lead to a greater appreciation for the complexity of political issues and the validity of diverse perspectives.

Encourage Compromise: Empathy can facilitate compromise by encouraging individuals to seek solutions that address the needs and concerns of all parties involved. When people feel understood and respected, they are more likely to be open to negotiation and compromise.

Empathy in Action: Examples and Strategies

Political Leaders as Role Models: Leaders who demonstrate empathy in their communication style can set a positive example for others to follow. Nelson Mandela's ability to forgive his oppressors and work towards reconciliation in South Africa serves as a powerful testament to the transformative potential of empathy. Similarly, Jacinda Ardern, the Prime Minister of New Zealand, has been lauded for her compassionate leadership in the wake of the Christchurch mosque shootings and the COVID-19 pandemic.

Countering Negativity in Online Discussions: Online platforms are often breeding grounds for hostility and incivility. However, empathy can be employed to counter this negativity. Individuals can practice empathy by:

Taking a moment to consider the perspective of the person they are engaging with.
Responding to comments with respect and avoiding personal attacks.
Seeking to understand the underlying reasons for someone's beliefs.
Using "I" statements to express their own views and feelings.
Media Literacy and Critical Thinking: Media literacy initiatives can help individuals to critically evaluate information and avoid the pitfalls of echo chambers and misinformation. By developing critical thinking skills, people can be more discerning consumers of news and information, which can help to reduce the spread of divisive rhetoric.

Cultivating Empathy in Political Discourse

Promoting empathy in political discourse requires a multi-faceted approach:

Education: Empathy should be integrated into educational curricula at all levels. Students should be taught how to recognize and understand emotions, perspective-take, and engage in civil discourse.

Dialogue Programs: Dialogue programs that bring together individuals from across the political spectrum can foster empathy and understanding. These programs provide a safe space for people to share their experiences, listen to others, and find common ground.

Community-Based Initiatives: Community-based initiatives, such as interfaith dialogues and community service projects, can bring people together across political lines and foster a sense of shared purpose.

Conclusion

Empathy is essential for promoting respectful and constructive political discourse. By fostering understanding, reducing hostility, and encouraging compromise, empathy can help to bridge political divides and create a more civil and productive political environment. As political polarization continues to pose a threat to democratic societies, the cultivation of empathy becomes increasingly critical.

12.3 Empathy and Political Leadership: Leading with Compassion and Understanding

The Essence of Empathetic Leadership

Empathy is not merely a desirable trait for political leaders; it is an essential ingredient for effective governance. Leaders who possess empathy are better equipped to connect with diverse constituencies, understand their needs and concerns, and build consensus around shared goals. Empathetic leadership fosters trust, promotes cooperation, and ultimately leads to decisions that benefit the common good.

Key Dimensions of Empathetic Political Leadership

Connecting with Diverse Constituencies: Empathetic leaders possess the ability to understand and appreciate the experiences of people from all walks of life. They actively seek out diverse perspectives, listen attentively to concerns, and strive to represent the interests of all citizens, regardless of their background or political affiliation.

Understanding Needs and Concerns: Empathy enables leaders to grasp the underlying needs and concerns that drive public opinion. By understanding the hopes, fears, and aspirations of their constituents, leaders can develop policies that address real-world challenges and resonate with the lived experiences of the people they serve.

Building Consensus: Empathy facilitates consensus-building by encouraging leaders to seek common ground and bridge divides. Empathetic leaders prioritize collaboration and compromise, recognizing that effective solutions often emerge from a process of genuine dialogue and mutual understanding.

Exemplary Empathetic Leaders

History provides compelling examples of political leaders who have embodied empathy in their leadership style:

Nelson Mandela: Mandela's extraordinary capacity for empathy was instrumental in guiding South Africa's transition from apartheid to democracy. His willingness to forgive his oppressors and engage in dialogue with former adversaries paved the way for reconciliation and healing.

Jacinda Ardern: Prime Minister Ardern of New Zealand has garnered international acclaim for her compassionate leadership in response to national tragedies and crises. Her ability to connect with people on an emotional level and offer genuine support has earned her the trust and admiration of her constituents.

Barack Obama: President Obama's leadership was often characterized by his ability to connect with people from diverse backgrounds and inspire hope. His emphasis on empathy and understanding helped to foster a sense of unity and common purpose during his time in office.

Cultivating Empathy in Political Leaders

Empathy is not an innate quality; it can be cultivated and practiced. Political leaders can develop their empathic capacity through:

Active Listening: Leaders must prioritize active listening, paying close attention to what others are saying and seeking to understand their perspectives.

Perspective-Taking: Engaging in perspective-taking exercises can help leaders to step outside their own experiences and see the world through the eyes of others.

Emotional Intelligence: Developing emotional intelligence enables leaders to recognize and understand their own emotions and the emotions of others, facilitating more effective communication and relationship-building.

Self-Reflection: Regular self-reflection can help leaders to identify their own biases and assumptions, leading to greater self-awareness and empathy.

Conclusion

Empathy is a critical attribute for effective political leadership. Leaders who demonstrate empathy are better able to connect with their constituents, understand their needs, and build consensus around shared goals. By cultivating empathy, political leaders can foster trust, promote cooperation, and make decisions that benefit the common good. In an increasingly polarized world, empathetic leadership is more important than ever.

12.4 Overcoming Political Polarization Through Empathy: Strategies for Building Bridges

The Urgency of Bridging Divides

Political polarization poses a significant threat to the health and stability of democratic societies. When individuals retreat into echo chambers and demonize those who hold different views, the foundation for compromise and cooperation erodes. This can lead to gridlock, social unrest, and even violence. Overcoming political polarization is not merely a matter of civility; it is essential for the preservation of democratic values and the pursuit of the common good.

Fostering Empathy Across Political Divides

Building bridges across political divides requires a concerted effort to foster empathy and understanding. A variety of strategies can contribute to this goal:

Intergroup Contact: Research has consistently shown that positive contact between individuals from different groups can reduce prejudice and promote understanding. This principle applies to political divides as well. When people from different political backgrounds interact in meaningful ways, they are more likely to see each other as individuals with shared values and concerns, rather than simply as members of an opposing group. Examples of intergroup contact initiatives include:

Dialogue groups: Structured dialogue programs bring together individuals from across the political spectrum to

engage in facilitated conversations about challenging issues. These programs emphasize active listening, respectful communication, and the search for common ground.
Community service projects: Working together on community service projects can create opportunities for people from different political backgrounds to collaborate towards a shared goal. This shared experience can foster a sense of community and break down stereotypes.
Interfaith initiatives: Interfaith dialogues and collaborative projects can bring together people from different religious and political backgrounds, promoting understanding and respect across lines of difference.
Dialogue Programs: Beyond simple intergroup contact, structured dialogue programs can play a crucial role in fostering empathy and understanding. These programs often utilize trained facilitators to guide participants through a process of sharing their perspectives, listening to others, and exploring areas of agreement and disagreement. Effective dialogue programs create a safe space for open and honest communication, fostering a sense of trust and mutual respect.

Media Literacy Initiatives: In the age of misinformation and echo chambers, media literacy education is essential for fostering critical thinking and combating polarization. Media literacy initiatives teach individuals how to:

Critically evaluate information sources.
Identify bias and propaganda.
Distinguish between facts and opinions.
Engage in responsible information sharing.
By equipping individuals with these skills, media literacy education can empower them to make informed decisions and resist the influence of divisive rhetoric.
Education: Education plays a foundational role in fostering empathy and promoting civic engagement.

From early childhood through higher education, curricula should emphasize:

Social and emotional learning: Developing social and emotional skills, such as empathy, self-awareness, and conflict resolution, is crucial for navigating social interactions and building healthy relationships.
Civic education: A strong foundation in civics equips individuals with the knowledge and skills necessary to participate effectively in democratic processes and engage in informed political discourse.
Critical thinking: Cultivating critical thinking skills enables individuals to analyze information, evaluate arguments, and form their own informed opinions.
Examples of Community-Based Initiatives

Living Room Conversations: This national organization facilitates conversations across divides by providing resources and support for individuals to host small group discussions in their homes or communities.

The People's Supper: This initiative brings people together for potluck meals and conversations about important issues, fostering connection and understanding across differences.

Essential Partners: This organization offers workshops and training programs that equip individuals with the skills to engage in constructive dialogue across divides.

Conclusion

Overcoming political polarization requires a multi-faceted approach that prioritizes empathy, understanding, and constructive engagement. By fostering intergroup contact, promoting dialogue, enhancing media literacy, and strengthening education, societies can build bridges across political divides and create a more cohesive and collaborative future. In a world increasingly marked by division and conflict, the pursuit of empathy and understanding is not merely a noble aspiration; it is an urgent necessity.

13.1 The Psychology of Narrative Empathy: Stepping into Another's Shoes Through Story

Delving into the Cognitive and Emotional Processes

Narrative empathy, the ability to understand and share the feelings of characters in a story, is a complex process involving both cognitive and emotional mechanisms. When we engage with a narrative, we step into the shoes of the characters, experiencing their world through their eyes. This immersive experience triggers a range of psychological processes that contribute to our empathetic engagement.

Mirror Neuron System and Emotional Contagion

One key mechanism underlying narrative empathy is the activation of the mirror neuron system. Mirror neurons are specialized brain cells that fire both when we perform an action and when we observe someone else performing the same action. This mirroring mechanism allows us to simulate the experiences of others, contributing to our understanding of their actions and emotions. When we read about a character experiencing joy or sadness, our mirror neurons fire as if we were experiencing those emotions ourselves, creating a sense of shared feeling.

Emotional contagion, another important process, refers to the phenomenon of "catching" the emotions of others. When we witness someone expressing an emotion, we tend to unconsciously mimic their facial expressions, body language, and even physiological responses. This mimicry leads to a convergence of emotional states, allowing us to feel what others are feeling. In the context of narratives, emotional contagion allows us to resonate with the emotional experiences of characters, further deepening our empathetic connection.

Theory of Mind and Social Cognition

Theory of mind, the ability to attribute mental states such as beliefs, desires, and intentions to oneself and others, plays a crucial role in narrative empathy. To understand a character's actions and emotions, we need to infer their underlying thoughts and motivations. This mentalizing process engages brain regions associated with theory of mind, such as the medial prefrontal cortex and the temporoparietal junction.

Social cognition, the broader set of processes involved in understanding and interacting with others, also contributes to narrative empathy. Narratives provide a rich source of information about social dynamics, relationships, and cultural norms. By observing how characters navigate social situations and interact with each other, we gain insights into human behavior and social relationships, enhancing our social cognitive abilities.

Research Evidence and Examples

Neuroimaging studies using fMRI have provided compelling evidence for the neural underpinnings of narrative empathy. These studies have shown that reading or watching stories activates brain regions associated with empathy, such as the anterior insula and the anterior cingulate cortex. These regions are involved in processing emotions, pain perception, and social cognition.

For example, a study by Speer et al. (2009) found that when participants read stories about characters experiencing pain, the same brain regions were activated as when they experienced pain themselves. This finding suggests that we simulate the experiences of characters at a neural level, blurring the lines between self and other.

Different narrative techniques can also modulate our empathetic engagement. First-person narration, by placing us directly in the mind of the character, can create a strong sense of identification and immersion. Close-up camera angles in films can similarly draw us into the emotional world of the characters, allowing us to witness their subtle facial expressions and emotional nuances.

Conclusion

Narrative empathy is a multifaceted process that draws on a range of cognitive and emotional mechanisms. By engaging our mirror neuron system, triggering emotional contagion, and activating brain regions associated with theory of mind and social cognition, narratives allow us to step into the shoes of others and experience the world from their perspective. This immersive experience can have a profound impact on our understanding of ourselves and others, fostering empathy, compassion, and social connection.

13.2 Empathy and Different Narrative Forms: Exploring the Power of Novels, Films, and Games

A Comparative Analysis of Narrative Forms and Their Empathetic Potential

Different narrative forms, such as novels, films, video games, and oral storytelling, offer unique affordances for evoking empathy and promoting understanding. Each medium employs distinct techniques to convey emotions, perspectives, and complex social realities, shaping our empathetic engagement in different ways.

Novels: Depth of Inner Worlds and Rich Descriptions

Novels excel at delving into the inner worlds of characters, providing access to their thoughts, feelings, and motivations through internal monologue, stream of consciousness, and rich descriptions. This deep dive into the psychological landscape of characters allows us to develop a profound understanding of their experiences, fostering empathy and identification.

For instance, in Fyodor Dostoevsky's Crime and Punishment, we are privy to the tormented mind of Rodion Raskolnikov, the protagonist, as he grapples with the guilt and paranoia following his crime. The novel's detailed exploration of Raskolnikov's psychological turmoil allows us to empathize with his struggles, even as we condemn his actions.

Films: Visual Cues, Music, and Emotional Resonance

Films, with their visual and auditory richness, can evoke powerful emotional responses through the use of visual cues, music, and sound design. Close-up shots, camera angles, and lighting can convey subtle emotional nuances, while music can amplify the emotional impact of scenes, creating a visceral sense of empathy.

Consider the film Schindler's List, which tells the story of Oskar Schindler, a German businessman who saved the lives of Jewish refugees during the Holocaust. The film's stark black and white cinematography, haunting music, and powerful performances create an emotionally charged experience, immersing viewers in the horrors of the Holocaust and fostering empathy for the victims.

Video Games: Immersive Experiences and Interactive Storytelling

Video games offer a unique opportunity for immersive and interactive storytelling, allowing players to step into the shoes of characters and experience the world through their eyes. By making choices and facing consequences within the game world, players can develop a deeper understanding of the characters' motivations and the complexities of their situations.

Games like Papers, Please, where players assume the role of an immigration officer in a fictional totalitarian state, challenge players to make difficult moral choices with real consequences. This interactive experience can foster empathy for individuals facing systemic oppression and spark reflection on ethical dilemmas.

Oral Storytelling: Shared Experience and Collective Memory

Oral storytelling, the oldest form of narrative transmission, relies on the power of the spoken word and the shared presence of storyteller and audience. Through vocal inflections, gestures, and facial expressions, storytellers bring stories to life, creating a sense of intimacy and connection. Oral traditions often carry cultural and historical knowledge, preserving collective memory and fostering a sense of community.

Indigenous cultures around the world have rich oral storytelling traditions that transmit knowledge, values, and history across generations. These stories often feature characters who interact with the natural world and embody cultural values, fostering empathy for both human and non-human beings.

Conclusion

Each narrative form offers unique affordances for evoking empathy and promoting understanding. Novels delve into the depths of inner worlds, films create visceral emotional experiences, video games offer immersive and interactive storytelling, and oral traditions foster shared experiences and preserve collective memory. By engaging with diverse narrative forms, we can expand our capacity for empathy and gain a deeper understanding of the human condition.

13.3 Storytelling for Social Change: Harnessing the Power of Narrative to Inspire Action

The Transformative Power of Storytelling

Storytelling has long been recognized as a powerful tool for social change. It has the unique ability to transcend barriers, connect with people on an emotional level, and inspire action. By sharing personal experiences, challenging stereotypes, and raising awareness of social issues, stories can mobilize communities and drive social progress.

Raising Awareness and Challenging Stereotypes

Stories can be instrumental in raising awareness of critical social issues that might otherwise remain hidden or misunderstood. By giving voice to marginalized communities and shedding light on injustice, stories can challenge prevailing narratives and spark conversations that lead to change.

#MeToo Movement: The #MeToo movement, which gained momentum in 2017, exemplifies the power of personal narratives to expose systemic issues and galvanize action. Through sharing their stories of sexual harassment and assault, individuals created a collective wave of awareness, challenging the silence and stigma surrounding these experiences. This surge of storytelling empowered others to come forward, leading to greater accountability and a shift in societal attitudes.

Black Lives Matter: The Black Lives Matter movement has effectively used storytelling to highlight the systemic racism and police brutality faced by Black communities. By sharing the stories of victims of police violence and amplifying the voices of Black activists, the movement has brought these issues to the forefront of public consciousness, sparking protests and demanding systemic change.

Humanizing Marginalized Groups and Inspiring Empathy

Stories have the power to humanize individuals and groups who have been marginalized and dehumanized by society. By sharing their stories, we can challenge stereotypes, foster empathy, and build bridges of understanding across social divides.

Refugee Narratives: Stories of refugees fleeing conflict and persecution can challenge the dehumanizing narratives often portrayed in the media. By sharing their personal journeys and struggles, refugees can reclaim their agency and foster empathy among those who may have preconceived notions about them. Organizations like Humans of New York have used storytelling to give a platform to refugees, allowing them to share their experiences in their own words and connect with a wider audience.

Disability Rights: Stories by and about people with disabilities can challenge stereotypes and misconceptions, promoting understanding and inclusion. By sharing their experiences of navigating a world that is often not designed for them, individuals with disabilities can raise awareness of the barriers they face and advocate for greater accessibility and equality.

Inspiring Action and Mobilizing Communities

Stories can be a catalyst for social change, inspiring individuals to take action and mobilize communities around a common cause. By connecting with people on an emotional level and providing a vision for a better future, stories can ignite passion and drive collective action.

Environmental Activism: Stories about the impact of climate change on communities and ecosystems can inspire individuals to take action to protect the environment. By sharing personal experiences of environmental degradation and highlighting the urgency of the crisis, activists can mobilize communities to demand change and engage in sustainable practices.

LGBTQ+ Rights: Stories of LGBTQ+ individuals and their struggles for equality have been instrumental in advancing LGBTQ+ rights. By sharing their experiences of discrimination, resilience, and love, LGBTQ+ individuals and allies have challenged prejudice, fostered understanding, and inspired action to create a more inclusive society.

Conclusion

Storytelling is a powerful tool for social change, capable of raising awareness, challenging stereotypes, humanizing marginalized groups, and inspiring action. By harnessing the power of narrative, we can build bridges of understanding, mobilize communities, and create a more just and equitable world.

13.4 Developing Empathy Through Creative Writing: Unlocking Personal Narratives and Perspectives

The Power of the Pen: Exploring the Self and Others Through Writing

Creative writing offers a powerful pathway to developing empathy and self-reflection. By engaging in the act of writing, we tap into our own emotions, explore different perspectives, and connect with others on a deeper level. Writing allows us to step outside of ourselves, to inhabit the minds and hearts of others, and to gain a richer understanding of the human experience.

Understanding Our Own Emotions Through Writing

Writing can serve as a therapeutic tool for self-discovery and emotional processing. By putting our thoughts and feelings into words, we gain clarity and insight into our own internal landscape. Journaling, poetry writing, and free writing exercises can help us to identify and express our emotions, leading to greater self-awareness and emotional regulation.

For example, writing about a challenging experience, such as a loss or a conflict, can help us to process our grief, anger, or confusion. By articulating our emotions, we can begin to make sense of them and move towards healing.

Exploring Different Perspectives Through Writing

Creative writing challenges us to step outside of our own limited perspectives and to imagine the world through the eyes of others. By writing from the point of view of a different character, we can gain a deeper understanding of their motivations, struggles, and experiences. This act of perspective-taking can broaden our empathy and challenge our assumptions about others.

Imagine writing a story from the perspective of a homeless person, a refugee, or someone from a different cultural background. This exercise can help us to appreciate the complexities of their lives and to challenge our own biases and prejudices.

Connecting with Others Through Writing

Writing can be a powerful tool for connection and communication. By sharing our stories with others, we can create a sense of shared experience and understanding. Writing can also help us to bridge cultural divides and to build empathy for those who are different from us.

Consider writing a letter to someone you have wronged or someone you have forgiven. This act of writing can be a powerful way to express your emotions, to seek reconciliation, or to offer forgiveness.

Prompts and Exercises for Developing Empathy Through Creative Writing

Here are some prompts and exercises that can help you to develop empathy through creative writing:

Write from the point of view of a character who is different from you in some significant way. This could be someone of a different age, gender, race, ethnicity, sexual orientation, or socioeconomic background.

Explore a challenging experience from multiple perspectives. Write about the same event from the point of view of different characters involved.

Write a letter of forgiveness to someone who has hurt you. This could be someone you know personally or a public figure.

Write a story about a time when you felt misunderstood. How did it feel? What did you learn from the experience?

Write a poem about a social issue that you care about. How does this issue affect you and others?

Create a fictional character who is struggling with a difficult emotion, such as grief, anger, or fear. How does this character cope with their emotions? What can you learn from their experience?

Conclusion

Creative writing is a powerful tool for developing empathy and self-reflection. By engaging in the act of writing, we can unlock our own personal narratives, explore different perspectives, and connect with others on a deeper level. Through writing, we can cultivate our capacity for empathy, compassion, and understanding, leading to a richer and more meaningful life.

Empathy and Mindfulness

14.1 Mindfulness and Emotional Awareness: Tuning in to Ourselves and Others
Exploring the Connection Between Mindfulness and Emotional Awareness

In the hustle and bustle of our daily lives, it's easy to become disconnected from our emotions. We may find ourselves reacting impulsively, feeling overwhelmed by stress, or struggling to understand the emotions of those around us. This is where mindfulness comes in. Mindfulness, the practice of paying attention to the present moment without judgment, can be a powerful tool for cultivating emotional awareness, both in ourselves and others.

How Mindfulness Increases Emotional Awareness

Mindfulness practices encourage us to turn our attention inward, observing our thoughts, feelings, and bodily sensations without judgment. This non-reactive observation allows us to become more aware of our emotional landscape, recognizing subtle shifts in our mood, identifying triggers, and understanding the underlying causes of our emotions.

By developing a deeper understanding of our own emotions, we can start to regulate our emotional responses more effectively. Instead of reacting impulsively, we can pause, observe our emotions, and choose how we want to respond. This increased emotional regulation can help us to navigate challenging situations with greater composure and resilience.

Furthermore, mindfulness can enhance our ability to empathize with others. By cultivating a non-judgmental awareness of our own emotions, we become more attuned to the emotional experiences of those around us. We can better understand their perspectives, recognize their emotional cues, and respond with greater sensitivity and compassion.

Mindfulness Practices for Cultivating Emotional Awareness

Here are a few mindfulness practices that can help you to develop greater emotional awareness:

Breath Awareness Meditation: This simple practice involves focusing your attention on the sensation of your breath as it enters and leaves your body. As you focus on your breath, notice any thoughts, feelings, or sensations that arise, acknowledging them without judgment and gently redirecting your attention back to your breath.
Body Scan Meditation: In this practice, you systematically bring your attention to different parts of your body, noticing any sensations, tension, or discomfort. This practice can help you to become more aware of your physical and emotional state, identifying areas of tension or discomfort that may be related to underlying emotions.
Loving-Kindness Meditation: This practice involves cultivating feelings of loving-kindness and compassion towards yourself and others. You can start by directing loving-kindness towards yourself, then gradually extend it to loved ones, acquaintances, strangers, and even those who have caused you harm. This practice can help you to develop a more compassionate and understanding attitude towards yourself and others.
Examples of Mindfulness in Action

Imagine you're in a meeting and a colleague criticizes your work. Instead of reacting defensively, you take a moment to pause and observe your emotions. You notice feelings of anger and frustration arising, but you also recognize that these emotions are fueled by a fear of failure. By acknowledging these emotions without judgment, you can respond to your colleague in a calm and constructive manner, addressing their concerns without letting your emotions get the better of you. Consider a scenario where a friend is going through a difficult time. By practicing mindfulness, you can become more attuned to their emotional state, recognizing their sadness, anxiety, or fear. This awareness allows you to offer support and compassion, creating a safe space for them to share their feelings and feel heard.

Research-Backed Benefits of Mindfulness

Numerous studies have demonstrated the benefits of mindfulness for emotional regulation and empathy. For instance, research has shown that mindfulness practices can:

Reduce emotional reactivity and increase emotional regulation
Enhance empathy and compassion
Improve communication skills and interpersonal relationships
Decrease stress, anxiety, and depression

Conclusion

Mindfulness is a powerful tool for cultivating emotional awareness, both in ourselves and others. By practicing mindfulness, we can develop a deeper understanding of our emotions, regulate our emotional responses, and cultivate greater empathy and compassion. This increased emotional intelligence can lead to more fulfilling relationships, improved communication, and greater well-being.

14.2 Mindfulness and Non-Judgmental Observation: Cultivating an Open and Accepting Mind

The Power of Non-Judgmental Observation

Our minds are constantly churning out thoughts, judgments, and evaluations. We judge ourselves, we judge others, and we judge our experiences. This constant stream of judgment can create a sense of inner turmoil, leading to stress, anxiety, and defensiveness. Mindfulness offers a powerful antidote to this judgmental mindset by cultivating non-judgmental observation.

Non-judgmental observation involves paying attention to our thoughts, feelings, and sensations without labeling them as good or bad, right or wrong. It's about observing our inner experience with curiosity and openness, accepting each moment as it is without trying to change it or push it away.

This practice of non-judgmental observation can have a profound impact on our well-being. When we stop judging ourselves and our experiences, we create space for greater self-acceptance and compassion. We can approach challenging situations with greater equanimity, responding to difficulties with wisdom and clarity rather than reactivity and defensiveness.

Furthermore, non-judgmental observation can enhance our empathy for others. When we can observe our own thoughts and feelings without judgment, we become more accepting of the thoughts and feelings of others. We can see beyond their behaviors and understand their perspectives, fostering deeper connection and understanding.

Mindfulness Exercises for Cultivating Non-Judgmental Awareness

Here are a few mindfulness exercises that can help you cultivate non-judgmental awareness:

Observing Thoughts Like Clouds: Imagine your thoughts as clouds passing across the sky. Observe each thought as it arises, noticing its shape, color, and movement. Don't try to hold onto any particular thought or push it away. Simply observe it with curiosity and allow it to pass.
Leaves on a Stream: Visualize your thoughts as leaves floating on a stream. Watch as each thought arises, floats along the stream, and eventually disappears downstream. Notice how the thoughts come and go without any effort on your part.
Mindful Walking: As you walk, pay attention to the sensations of your feet on the ground, the movement of your body, and the sights and sounds around you. Notice any thoughts or judgments that arise, acknowledging them without judgment and gently redirecting your attention back to your sensory experience.
Examples of Non-Judgmental Observation in Daily Life

Imagine you make a mistake at work. Instead of berating yourself with self-criticism, you observe the thought, "I made a mistake," without judgment. You acknowledge the mistake, learn from it, and move on, without letting it define your worth or your abilities.
Consider a situation where you're having a disagreement with a loved one. Instead of judging their perspective or getting defensive, you practice non-judgmental observation. You listen to their point of view with an open mind, seeking to understand their perspective without judgment. This approach can help to de-escalate conflict and foster a more constructive dialogue.

The Science Behind Non-Judgmental Observation

Research suggests that non-judgmental observation can have a number of positive effects on our mental and emotional health. For example, studies have shown that mindfulness practices can:

Reduce rumination and negative thinking
Decrease stress, anxiety, and depression
Improve emotional regulation and resilience
Enhance self-compassion and acceptance

Conclusion

Non-judgmental observation is a cornerstone of mindfulness practice. By cultivating an open and accepting mind, we can break free from the cycle of judgment and create space for greater self-awareness, compassion, and empathy. This shift in perspective can lead to more fulfilling relationships, improved emotional well-being, and a greater sense of inner peace.

14.3 Mindfulness and Compassion: Extending Kindness to Ourselves and Others

The Intertwined Nature of Mindfulness and Compassion

Mindfulness and compassion are deeply intertwined. Mindfulness, with its emphasis on present moment awareness and non-judgmental observation, creates fertile ground for the cultivation of compassion. By becoming more aware of our own suffering and the suffering of others, we naturally develop a greater capacity for empathy and kindness.

Compassion involves recognizing the suffering of ourselves and others and feeling motivated to alleviate that suffering. It's about offering kindness, understanding, and support, both to ourselves and those around us.

Compassion is not pity or sympathy; it's an active engagement with suffering, a desire to help, and a recognition of our shared humanity.

Mindfulness practices can help us to cultivate compassion in several ways. By developing a deeper awareness of our own emotions and experiences, we become more attuned to the suffering within ourselves. This self-awareness can lead to greater self-compassion, allowing us to treat ourselves with kindness and understanding, especially during times of difficulty.

Furthermore, mindfulness can enhance our empathy for others. By cultivating a non-judgmental awareness of our own emotions, we become more sensitive to the emotional experiences of those around us. We can better understand their perspectives, recognize their pain, and respond with compassion and care.

Mindfulness Practices for Cultivating Compassion

Here are a few mindfulness practices that can help you cultivate compassion:

Tonglen Meditation: This Tibetan Buddhist practice involves visualizing taking in the suffering of others with each inhalation and sending out compassion and loving-kindness with each exhalation. This practice can help us to connect with the suffering of others and develop a deeper sense of empathy.
Loving-Kindness Meditation: This practice involves cultivating feelings of loving-kindness and compassion towards oneself and others. You can start by directing loving-kindness towards yourself, then gradually extend it to loved ones, acquaintances, strangers, and even those who have caused you harm. This practice can help you to

develop a more compassionate and understanding attitude towards yourself and others.
Compassionate Body Scan: This practice combines the body scan meditation with compassion. As you scan your body, notice any areas of tension, discomfort, or pain. Instead of judging these sensations, offer yourself compassion and kindness, acknowledging your suffering and wishing yourself well. You can then extend this compassion to others, visualizing them experiencing similar sensations and offering them loving-kindness.
Examples of Compassion in Action

Imagine a friend is going through a breakup. Instead of offering advice or trying to fix the situation, you simply listen with compassion, offering your presence and support. You acknowledge their pain, validate their feelings, and offer words of comfort and encouragement.
Consider a situation where you witness someone experiencing homelessness. Instead of judging or turning away, you feel a sense of compassion for their suffering. You may offer them food, money, or simply a kind word, recognizing their humanity and offering support in whatever way you can.
The Science of Compassion

Research has shown that compassion has numerous benefits for both our mental and physical health. For example, studies have found that compassion can:

Increase happiness, well-being, and life satisfaction
Reduce stress, anxiety, and depression
Strengthen the immune system and improve physical health Enhance social connection and relationships
Conclusion

Mindfulness and compassion are powerful tools for creating a more kind and compassionate world. By cultivating present moment awareness and non-judgmental observation, we can develop a deeper understanding of our own suffering and the suffering of others. This understanding can lead to greater self-compassion, empathy, and kindness, allowing us to connect with our shared humanity and create a more compassionate and caring world.

14.4 Mindfulness and Interpersonal Relationships: Communicating with Presence and Understanding

Mindfulness as a Foundation for Healthy Relationships

Our relationships are at the heart of our lives, providing us with love, support, and connection. Yet, navigating the complexities of interpersonal dynamics can be challenging. Miscommunication, misunderstandings, and conflicts can strain even the strongest bonds. This is where mindfulness can play a transformative role.

By cultivating present moment awareness and non-judgmental observation, mindfulness can enhance our communication skills, deepen our empathy, and foster more fulfilling relationships. It allows us to be more present in our interactions, listen more attentively, and respond with greater compassion and understanding.

How Mindfulness Improves Communication

Mindfulness can improve communication in several ways:

Active Listening: Mindfulness encourages us to truly listen to what others are saying, not just waiting for our turn to speak. We pay attention to their words, tone of voice, and body language, seeking to understand their perspective fully. This deep listening creates a sense of being heard and understood, fostering trust and connection.

Mindful Speaking: Mindfulness can also help us to communicate more effectively. We become more aware of our own thoughts and emotions, allowing us to express ourselves with clarity and authenticity. We choose our words carefully, considering their impact on others, and speak with kindness and compassion.
Non-Verbal Communication: Mindfulness enhances our awareness of non-verbal cues, both in ourselves and others. We become more attuned to body language, facial expressions, and subtle shifts in energy, allowing us to communicate more effectively and respond to others with greater sensitivity.
Mindfulness Practices for Enhancing Communication

Here are a few mindfulness exercises that can help you enhance your communication skills:

Mindful Listening: Practice listening to others with your full attention. Put aside distractions, make eye contact, and focus on understanding their message. Notice any judgments or assumptions that arise, and gently redirect your attention back to their words.
Mindful Speaking: Before you speak, take a moment to pause and connect with your breath. Consider what you want to say and how you want to say it. Choose your words carefully, speaking with clarity, kindness, and compassion.
Mindful Conflict Resolution: When conflicts arise, approach the situation with mindfulness. Acknowledge your own emotions and seek to understand the perspective of the other person. Communicate your needs and concerns clearly and respectfully, and be willing to listen to their perspective.
Examples of Mindfulness in Relationships

Imagine you're having a conversation with your partner. Instead of interrupting or thinking about what you want to say next, you practice mindful listening. You give your partner your full attention, listening to their words and observing their body language. This deep listening allows you to understand their perspective fully, fostering a deeper connection and understanding.

Consider a situation where you're feeling stressed and overwhelmed. Instead of taking your frustration out on your loved ones, you practice mindfulness. You take a moment to pause, connect with your breath, and acknowledge your emotions. This awareness allows you to communicate your needs clearly and calmly, without resorting to anger or blame.

The Impact of Mindfulness on Relationships

Research has shown that mindfulness can have a profound impact on our relationships. For example, studies have found that mindfulness practices can:

- Increase empathy and compassion
- Improve communication skills and conflict resolution
- Enhance intimacy and connection
- Reduce stress and conflict in relationships
- Foster greater relationship satisfaction

Conclusion

Mindfulness is a powerful tool for cultivating healthy and fulfilling relationships. By bringing present moment awareness and non-judgmental observation into our interactions, we can communicate more effectively, deepen our empathy, and foster greater connection and understanding. This mindful approach to relationships can lead to greater intimacy, compassion, and support, creating stronger and more resilient bonds.

The Limits of Empathy

15.1 Empathy Fatigue: Recognizing and Preventing Burnout

Empathy fatigue, also known as compassion fatigue or secondary traumatic stress, is a state of emotional and physical exhaustion that can result from prolonged exposure to the suffering or trauma of others. It is often characterized by feelings of overwhelm, cynicism, detachment, and a reduced capacity for empathy. This condition is particularly prevalent in helping professions such as healthcare, social work, and education, where individuals are frequently exposed to the distress and trauma of those they serve.

Signs and Symptoms of Empathy Fatigue

Recognizing the signs and symptoms of empathy fatigue is crucial for early intervention and prevention. These may include:

Emotional exhaustion: Feeling emotionally drained, depleted, and unable to cope with the emotional demands of your work or personal life.
Physical fatigue: Experiencing persistent physical tiredness, sleep disturbances, and a weakened immune system.
Reduced empathy: Feeling less able to connect with and understand the emotions of others, leading to detachment and cynicism.
Increased irritability and negativity: Becoming easily frustrated, irritable, and having a more negative outlook on life.
Difficulty concentrating: Experiencing difficulty focusing, making decisions, and remembering things.

Feeling overwhelmed and helpless: Feeling a sense of hopelessness and helplessness in the face of others' suffering.
Increased use of substances or unhealthy coping mechanisms: Turning to alcohol, drugs, or other unhealthy behaviors to cope with stress.
Strategies for Preventing and Managing Empathy Fatigue

Preventing and managing empathy fatigue requires a proactive and multifaceted approach that prioritizes self-care, stress management, and healthy boundaries. Here are some strategies that can help:

Setting boundaries: Establishing clear boundaries between your personal and professional life is essential for protecting your emotional well-being. This may involve setting limits on your work hours, taking regular breaks, and learning to say no to additional responsibilities when you are feeling overwhelmed.
Practicing mindfulness: Mindfulness techniques, such as meditation and deep breathing exercises, can help you stay grounded in the present moment and reduce feelings of stress and anxiety. Regular mindfulness practice can also enhance your self-awareness and help you recognize the early signs of empathy fatigue.
Seeking support from colleagues and supervisors: Sharing your experiences and concerns with trusted colleagues and supervisors can provide emotional support and validation. Peer support groups and professional supervision can also offer a safe space to process challenging emotions and develop coping strategies.
Engaging in self-care activities: Prioritizing self-care is crucial for maintaining your emotional and physical well-being. This may include activities such as exercise, spending time in nature, pursuing hobbies, and connecting with loved ones.

Taking regular breaks and vacations: Taking time away from work to rest and recharge is essential for preventing burnout. Make sure to schedule regular breaks throughout your day and take vacations to allow yourself time to recover from the emotional demands of your work.
Seeking professional help: If you are struggling with empathy fatigue, seeking professional help from a therapist or counselor can provide you with additional support and guidance. Therapy can help you develop coping strategies, process challenging emotions, and build resilience.

Examples and Case Studies

A study published in the Journal of Advanced Nursing in 2023 found that nurses working in intensive care units had a significantly higher prevalence of empathy fatigue compared to nurses working in other settings. The study also found that nurses who reported higher levels of perceived stress and lower levels of social support were more likely to experience empathy fatigue.
A case study published in the Clinical Social Work Journal in 2024 described the experience of a social worker who developed empathy fatigue after working with refugees who had experienced significant trauma. The social worker reported feeling emotionally exhausted, detached from her clients, and having difficulty sleeping. Through therapy and self-care practices, she was able to recover from empathy fatigue and return to her work with renewed energy and compassion.

Experts Insights

Dr. Charles Figley, a leading expert on compassion fatigue, emphasizes the importance of recognizing the early signs of empathy fatigue and taking proactive steps to prevent burnout. He recommends that individuals in helping professions prioritize self-care, set healthy boundaries, and seek support from colleagues and supervisors.

Dr. Kristin Neff, a renowned researcher on self-compassion, suggests that practicing self-compassion can be a powerful tool for managing empathy fatigue. She encourages individuals to treat themselves with the same kindness and understanding that they would offer to a friend who is struggling.

15.2 Empathy and Bias: Understanding the Limits of Our Compassion

Empathy, while a powerful force for good, is not without its limitations. Our capacity for empathy can be significantly influenced by various biases, including group membership, similarity, and perceived deservingness. These biases can shape our empathetic responses, leading to unfair or discriminatory behavior.

In-group/Out-group Bias

One of the most well-established biases in social psychology is the in-group/out-group bias. This bias refers to our tendency to favor members of our own group (in-group) over those who belong to different groups (out-group). This bias can manifest in various ways, including:

Greater empathy for in-group members: We are more likely to feel empathy for those who share our social identities, such as race, ethnicity, gender, or nationality. This can lead to a disparity in our empathetic responses, where we feel more compassion for the suffering of in-group members compared to out-group members.
Dehumanization of out-group members: In extreme cases, the in-group/out-group bias can lead to the dehumanization of out-group members, where they are

perceived as less human or deserving of empathy. This can have devastating consequences, as it can justify discrimination, prejudice, and even violence against out-group members.

Similarity Bias

Our empathy can also be influenced by our perceived similarity to others. We tend to feel more empathy for those who are similar to us in terms of their values, beliefs, experiences, or even physical appearance. This bias can lead to:

Selective empathy: We may be more likely to extend empathy to those who are similar to us, while neglecting the suffering of those who are different. This can create a cycle of empathy where certain groups are consistently marginalized and denied compassion.

Reinforcement of existing social inequalities: The similarity bias can perpetuate existing social inequalities by favoring those who already have privilege and power. This can make it more difficult for marginalized groups to gain access to resources and support.

Perceived Deservingness

Our empathy can also be influenced by our perceptions of whether someone deserves our compassion. This can be based on various factors, including:

Attributions of responsibility: We may be less likely to feel empathy for those who we believe are responsible for their own suffering. This can lead to a "blame the victim" mentality, where individuals are held accountable for circumstances beyond their control.

Moral judgments: Our moral judgments about someone's behavior can also influence our empathy. We may be less likely to feel empathy for those who we believe have acted immorally or violated social norms.

Stereotypes and prejudice: Stereotypes and prejudice can distort our perceptions of others and lead to inaccurate assessments of their deservingness. This can result in discriminatory empathetic responses, where certain groups are denied compassion based on their social identities.

Examples and Case Studies

A study published in the Journal of Personality and Social Psychology in 2023 found that participants showed greater empathy for victims of a natural disaster when the victims were from their own country compared to when they were from a different country. This highlights the influence of the in-group/out-group bias on our empathetic responses.

A case study published in the journal Social Cognition in 2024 examined the role of perceived deservingness in empathy. The study found that participants were less likely to feel empathy for a homeless person who was perceived as being responsible for their own situation, compared to a homeless person who was perceived as being a victim of circumstances beyond their control.

Experts Insights

Dr. Brené Brown, a leading researcher on vulnerability and shame, emphasizes the importance of recognizing our own biases and how they can impact our empathy. She encourages individuals to cultivate a sense of curiosity and openness towards those who are different from them, and to challenge their own assumptions and stereotypes.

Dr. Paul Bloom, a professor of psychology and cognitive science, argues that empathy can be a limited and unreliable guide for moral decision-making. He suggests that we should rely on reason and principles of justice to guide our actions, rather than solely relying on our emotional responses.

15.3 Empathy and Moral Decision-Making: Balancing Compassion with Justice

Empathy plays a crucial role in moral decision-making, but it's not always a straightforward or reliable guide. While empathy can motivate us to act compassionately and help those in need, it can also lead to biased or suboptimal choices if not balanced with other ethical considerations, such as fairness, justice, and impartiality.

The Potential Pitfalls of Empathy in Moral Decision-Making

Emotional bias: Empathy can be heavily influenced by emotions, leading to decisions that prioritize the immediate needs of individuals we feel empathy for, even if those decisions are not fair or just in the broader context.
Focus on the individual: Empathy tends to focus on the suffering of specific individuals, potentially overlooking the needs of larger groups or the long-term consequences of our actions.
Vulnerability to manipulation: Individuals can exploit our empathy to gain favors or manipulate us into making decisions that benefit them, even if those decisions are unethical or harmful to others.
In-group bias: As discussed earlier, our empathy is often stronger for those we perceive as similar to us, leading to biased moral judgments and actions that favor in-group members over out-group members.

Balancing Empathy with Other Ethical Considerations

To make sound moral decisions, it's essential to balance empathy with other ethical principles, such as:

Fairness: Ensuring that our actions are equitable and do not unfairly disadvantage certain individuals or groups.

Justice: Considering the broader implications of our actions and ensuring they align with principles of justice and social responsibility.
Impartiality: Striving to make decisions based on objective criteria and avoiding favoritism or bias.
Utilitarianism: Weighing the potential benefits and harms of our actions for all parties involved and choosing the option that maximizes overall well-being.
Ethical Dilemmas Highlighting the Conflict Between Empathy and Other Moral Values

The Trolley Problem: A classic thought experiment in ethics, the trolley problem presents a scenario where a runaway trolley is about to kill five people. You have the option to divert the trolley to a different track, where it will kill one person instead. This dilemma highlights the conflict between empathy (feeling for the one person who would be sacrificed) and utilitarianism (saving the greater number of lives).
Resource Allocation in Healthcare: In situations where medical resources are scarce, healthcare professionals often face difficult decisions about how to allocate those resources. Empathy for individual patients may conflict with the need to prioritize those with the greatest chance of survival or to distribute resources fairly among the population.
Criminal Justice: In the criminal justice system, empathy for victims may conflict with the need to ensure fair trials and avoid excessive punishment. Balancing empathy with principles of justice and due process is crucial for maintaining a fair and equitable legal system.
Using Empathy to Inform Ethical Decision-Making

While empathy should not be the sole determinant of moral decisions, it can be a valuable tool for informing ethical reasoning. By cultivating empathy, we can:

Gain a deeper understanding of the perspectives and needs of others.
Identify potential harms and benefits of our actions.
Motivate ourselves to act compassionately and promote the well-being of others.
Examples and Case Studies

A study published in the journal Psychological Science in 2025 investigated the role of empathy in moral decision-making. The study found that participants who were induced to feel empathy for a victim were more likely to make decisions that favored that individual, even if those decisions were unfair to others.
A case study published in the Journal of Business Ethics in 2024 examined the ethical challenges faced by a company that was considering closing a factory in a small town. The company's leaders felt empathy for the employees who would lose their jobs, but they also had to consider the financial well-being of the company and its shareholders. The case study highlights the complex interplay between empathy, economic considerations, and social responsibility in ethical decision-making.
Experts Insights

Dr. Jonathan Haidt, a social psychologist and author of "The Righteous Mind," argues that moral decision-making is influenced by a variety of factors, including intuition, emotion, and reasoning. He emphasizes the importance of balancing empathy with other moral foundations, such as fairness, loyalty, authority, and sanctity.

Dr. Daniel Kahneman, a Nobel laureate and author of "Thinking, Fast and Slow," highlights the role of cognitive biases in decision-making. He suggests that we are often influenced by emotional heuristics and mental shortcuts, which can lead to suboptimal choices. Cultivating awareness of these biases can help us make more rational and ethical decisions.

15.4 Cultivating Ethical and Sustainable Empathy: A Path to Compassionate Action

Cultivating empathy is not simply about feeling more; it's about feeling more effectively. It's about developing a compassionate approach that is both ethical and sustainable, avoiding the pitfalls of burnout, bias, and emotional overwhelm. This requires a conscious effort to develop self-awareness, self-compassion, and critical reflection.

Key Principles for Cultivating Ethical and Sustainable Empathy

Self-awareness: Understanding your own emotional landscape, biases, and triggers is crucial for managing your empathetic responses and avoiding burnout. This involves paying attention to your own emotions, recognizing your limitations, and identifying situations that are particularly emotionally challenging for you.
Self-compassion: Treating yourself with kindness and understanding, especially when you are struggling or feeling overwhelmed, is essential for maintaining emotional resilience. Self-compassion involves recognizing that you are not perfect, accepting your own vulnerabilities, and offering yourself the same compassion you would offer to a friend.
Critical reflection: Regularly reflecting on your empathetic responses can help you identify biases, challenge your assumptions, and refine your approach to empathy. This involves asking yourself questions like: "Am I feeling more empathy for certain individuals or groups? Why? Are my empathetic responses aligned with my values and ethical principles?"

Boundaries: Setting healthy boundaries between your personal and professional life is crucial for protecting your emotional well-being and preventing empathy fatigue. This may involve limiting your exposure to traumatic material, taking breaks from emotionally demanding situations, and prioritizing your own self-care.
Perspective-taking: Actively seeking out diverse perspectives and challenging your own assumptions can help you overcome biases and cultivate a more inclusive and compassionate worldview. This involves engaging with people from different backgrounds, listening to their stories, and considering their experiences.
Mindfulness: Practicing mindfulness can help you stay grounded in the present moment, regulate your emotions, and cultivate a sense of calm and clarity. This can be particularly helpful in emotionally charged situations, allowing you to respond with greater compassion and wisdom.
Practical Tips for Cultivating Ethical Empathy

Practice active listening: Pay attention to what others are saying, both verbally and nonverbally, and strive to understand their perspectives.
Challenge your assumptions: Be willing to question your own beliefs and biases, and be open to learning from others.
Cultivate curiosity: Approach interactions with a sense of curiosity and openness, seeking to understand others' experiences and motivations.
Practice gratitude: Focusing on the positive aspects of your life and expressing gratitude can help you cultivate a more compassionate and optimistic outlook.
Engage in self-care: Prioritize activities that nourish your mind, body, and spirit, such as exercise, spending time in nature, pursuing hobbies, and connecting with loved ones.

Seek support: Don't be afraid to reach out to friends, family, colleagues, or a therapist for support when you are feeling overwhelmed or struggling with empathy fatigue.

Examples and Case Studies

A study published in the journal Mindfulness in 2025 found that participants who completed a mindfulness-based compassion training program showed significant increases in empathy, self-compassion, and emotional well-being.

A case study published in the Journal of Social Work Education in 2024 described the experience of a social work student who learned to cultivate ethical empathy through a combination of self-reflection, mindfulness practice, and engagement with diverse communities. The student reported that this process helped her to become a more compassionate and effective social worker.

Experts Insights

Dr. Dalai Lama, a spiritual leader and advocate for compassion, emphasizes the importance of cultivating both wisdom and compassion. He suggests that wisdom helps us to understand the nature of reality, while compassion motivates us to act with kindness and alleviate suffering.

Dr. Sharon Salzberg, a leading teacher of mindfulness and compassion, encourages individuals to cultivate a "lovingkindness" meditation practice, which involves directing feelings of warmth and goodwill towards oneself and others. This practice can help to cultivate a more compassionate and interconnected worldview.

Conclusion

Cultivating ethical and sustainable empathy is an ongoing process that requires self-awareness, self-compassion, and a commitment to personal growth. By embracing these principles and practicing the tips outlined above, we can develop a more compassionate and effective approach to engaging with the world around us. This not only benefits those we interact with but also enhances our own well-being and sense of purpose.

About Author

Azhar ul Haque Sario is a bestselling author and data scientist. A Cambridge alumnus, he possesses proven technical skills with an MBA, ACCA (Knowledge Level - FTMS College Malaysia), BBA, and several Google certifications, including Google Data Analytics Specialization, Google Digital Marketing & E-commerce Specialization, and Google Project Management Specialization.

With ten years of business experience, Azhar leverages his expertise to deliver insightful books. He has written and published an astounding 2810 titles and was awarded the record for the maximum Kindle editions and paperback books published by an individual author in one year by Asia Books of Records in 2024.

ORCID: https://orcid.org/0009-0004-8629-830X
Azhar.sario@hotmail.co.uk

www.ingramcontent.com/pod-product-compliance
Lightning Source LLC
Chambersburg PA
CBHW031348230926
51848CB00002B/436
9783384627360